D0856429

The Natural Selection of Populations and Communities

The Natural Selection of Populations and Communities

David Sloan Wilson
University of California, Davis

THE BENJAMIN/CUMMINGS PUBLISHING COMPANY, INC.
Menlo Park, California · Reading, Massachusetts
London · Amsterdam · Don Mills, Ontario · Sydney

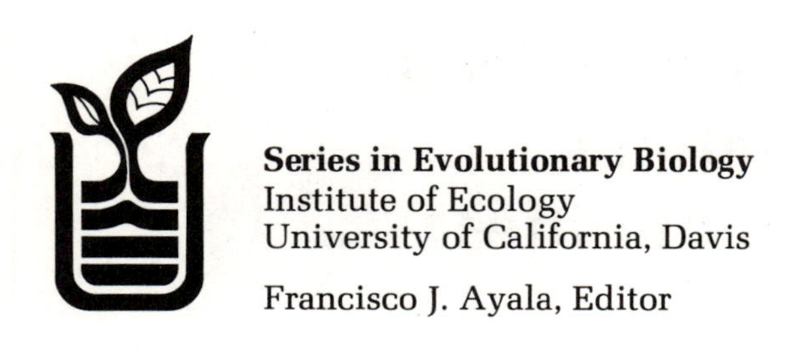

Book production by Greg Hubit Bookworks

Library of Congress Cataloging in Publication Data

Wilson, David Sloan.
The natural selection of populations and communities.

(Series in evolutionary biology)
Bibliography: p.
Includes index.
1. Natural selection. 2. Population genetics.
3. Biotic communities. I. Title. II. Series.
QH375.W54 575.01'62 79-19139
ISBN 0-8053-9560-1

abcdefghij-HA-8210

The Benjamin/Cummings Publishing Company, Inc.
2727 Sand Hill Road
Menlo Park, California 94025

To: Sloan and Betty,
Elise and Sawnie,
and Anne

Foreword

As Darwin saw it, natural selection is the fundamental process that accounts for the evolutionary change of organisms through the generations and for the adaptations of organisms. In outline, Darwin's argument is simple. Organisms with hereditary variations that make them well adapted to the environments in which they live are likely to leave more descendants than organisms with less adaptive variations. It follows that hereditary variations that enhance adaptation will gradually increase in frequency from generation to generation and eventually replace less adaptive alternative variations. Thus, natural selection is a process that explains the occurrence of evolution as well as why living organisms are adapted to their environments and ways of life.

Can natural selection account for all the sorts of adaptations observed in living beings? Natural selection may be defined as differential reproduction of alternative genetic variants. Natural selection would, then, account for any adaptations that increase the reproductive efficiency of their carriers relative to that of carriers of alternative properties. Most of the adaptations observed in organisms seem to meet this requirement, but not all adaptations do. Consider, for example, altruism; that is, forms of behavior that benefit other individuals at the expense of the altruist. A crow, sitting in a tree watching for predators while the rest of the flock forages, benefits the flock by providing sentinel protection, but it is incurring a loss by not feeding in the interim. The probability of reproduction of the altruist is decreased, whereas that of individuals that do not behave altruistically is enhanced by the altruist's behavior. If the behavior is genetically determined, it seems that natural selection will favor nonaltruistic alleles at the expense of altruistic ones.

Altruism is only one of various kinds of adaptation that may be beneficial to the population, but whose existence cannot simply be accounted for as the result of natural selection favoring some genes at the expense of alternative genes within a population. Some authors have proposed that adaptations that are useful to a population as a group may come about by *group selection*, a modified form of natural selection that operates as the result of the differential reproduction of populations or groups. The argument is that groups endowed with such adaptations are more likely to persist than groups lacking them.

Local populations do become extinct on occasion and may be reestablished by colonizers from other populations of the same species. If this were all that is claimed by the proponents of group selection, it would be a tame and, indeed, unobjectionable proposal. However, group selection has been advanced as an explanation of the occurrence of adaptations—such as altruism—that are beneficial to the population, but are not favored by natural selection within the population when alternative genetic constitutions exist. How could such adaptations become established in populations? Models have been proposed that take into account the possibility of occasional fixation of such adaptations by genetic drift, their disadvantage within the population, and the rate of extinction and recolonization of local populations. However, many biologists think that the conditions required are excessively restricted. These biologists discount group selection as a likely explanation for the alleged common occurrence of population adaptations that are not favored by natural selection at the individual level.

But are individual selection and group selection two mutually exclusive alternatives? David S. Wilson proposes in this book that individual and group selection should rather be seen as the two extremes of a continuum. Wilson starts by removing the "homogeneity assumption"—the simplifying assumption made in population genetics that genotypes are uniformly distributed within a population (deme). Wilson argues that uniform genotypic distribution is not always, and perhaps not generally, the case. Instead, populations are often *structured demes*, consisting of small subgroups, which Wilson calls *trait groups*, that may differ in their allelic frequencies.

Assume that trait groups possessing a certain allele produce greater numbers of progeny than trait groups lacking that allele (or having it at lower frequency). The frequency of the allele in the population might increase from generation to generation, even if the allele decreases in frequency within each trait group in which it is present. Whether or not the allele will increase in the population as a whole depends on various parameters, such as the genetic heterogeneity between trait groups, the redistribution of individuals among trait groups from one generation to another, the differential productivity of the trait groups, and the selective disadvantage of the allele within a trait group. Limiting values of these parameters define the two traditional cases presented earlier of individual selection (when the trait groups are genetically homogeneous) and group selection (when the groups are effectively autonomous, except that extinct trait groups are replaced by colonizers from persisting trait groups). A continuum of situations between the two extremes, as a function of the genetic variation between trait groups, is shown by Wilson in Figure 2.2 (page 28).

Wilson is aware of the assumptions and simplifications he has made in developing the theory of structured demes. But he argues that his models are likely to be robust and to have general application. He adduces empirical evidence to support critical elements of the theory. Most important, however, is that the theory can be subject to empirical testing. The ultimate role that the theory of structured demes will play in evolutionary science will be decided by its empirical corroboration or falsification. I predict that, at least for the next few years, the theory will be extensively discussed and tested by evolutionists.

□□□

The Natural Selection of Populations and Communities inaugurates the Series in Evolutionary Biology sponsored by the Institute of Ecology at the University of California, Davis. The objective of the Series is to publish important conceptual contributions to evolutionary thought. The Series will encompass the whole spectrum of evolutionary disciplines, from genetics and ecology to biogeography and paleontology, as well as all levels of organization, from the molecular through the organismal and populational to the community. The books will be produced by The Benjamin/Cummings Publishing Company.

Francisco J. Ayala
Davis, California

Preface

In 1962, V. C. Wynne-Edwards proposed that natural selection promotes the survival and productivity of whole populations in his monumental book, *Animal Dispersion in Relation to Social Behavior*. The idea has been hotly contested ever since. I was first attracted to the controversy as an undergraduate student in 1971 and took it up in earnest in 1974. This book represents what I hope to be a fresh approach to the group selection problem, and also extends the concepts to an examination of multispecies coevolution. If the models presented here are correct, then the conditions for the natural selection of populations and communities are not as stringent as sometimes believed.

Work on the book started at the University of Michigan Biological Station and continued at the University of Washington, the University of the Witwatersrand (Johannesburg), the Council for Scientific and Industrial Research (Pretoria), and the University of California at Davis. I am particularly grateful to the two South African institutions, where the bulk of the first draft was completed. Hugh Paterson and Barry Fabian gave me free office space and library privileges at the University of the Witwatersrand. David Jacobson gave me a salary at CSIR, even though my work bore no obvious relation to the rest of the institute's activities.

I bothered many people with questions and bulky first drafts. E. I. Charnov, D. Cohen, R. K. Colwell, P. H. Crowley, S. D. Fretwell,

S. Gleeson, W. D. Hamilton, C. A. Istock, J. C. Moser, S. Rohwer, T. W. Schoener, M. Slatkin, C. K. Starr, T. J. Stewart, M. Turelli, L. Van Valen, M. J. Wade, G. C. Williams, and E. O. Wilson read all or parts of various drafts. E. I. Charnov, A. B. Clark, R. K. Colwell, C. A. Istock, and M. Turelli were especially close consultants, and many of the book's best points are attributable to them. Anne Clark combined the roles of wife and devil's advocate with consummate skill.

In addition to his substantive comments, Rob Colwell provided a fine example of strong altruism by correcting all my grammatical errors. After seeing how many there were, I have decided to reenroll in the fifth grade during my next sabbatical. Lola Brocksen patiently and expertly typed two drafts. Ann Hedrick helped enormously with the page proofs. I am especially grateful to Shahid Naeem, who drew the cover illustration representing the hummingbird *Glaucis hirsuta* and its associated mites of the genus *Rhinoseius* (Mesostigmata: Ascidae). The mites use the hummingbird for transport, to get from flower to flower. The fine work of J. C. Moser and his collaborators, which I believe supports my models, was funded by the USDA Forest Service, Southern Forest Experiment Station. My own work has been funded by NSF grants BMS74-20550, BMS75-17663, and DEB78-03153.

Finally, I would like to thank Dr. V. C. Wynne-Edwards, not only for his generous advice and enthusiasm, but also for creating the controversy in the first place.

David Sloan Wilson
Davis, California

Contents

The Natural Selection of Populations and Communities

1 Introduction

Selfishness and Superorganisms

This book examines the relationship between an individual organism and the community to which it belongs. By community, I refer to other organisms, of both the same and different species, with which the individual interacts. The main question raised in the book can be stated as follows: Are we justified, as evolutionary biologists, in saying that an individual exists for the function it performs in its community? Are there any aspects of an individual's behavior and morphology that can be understood only in terms of its role in maintaining a larger adaptive unit?

The idea that an individual is a cell in a larger "superorganism" has a long history, but not the kind that can be used to support an evolutionary hypothesis. Ghiselin (1974) argues that this idea was the predominant pre-Darwinian paradigm of nature—a universe ordered by God, in which everything plays a meaningful part.

> *The basic feature of the Darwinian revolution has been a shift in our conception of the status of groups and individuals, and in the concepts of the relations between them. Much as Copernicus moved the sun to the center of the solar system, Darwin placed the organism at the center of the biological universe. The metaphor, indeed,*

aptly expresses the historical situation. To Plato and Aristotle the entire cosmos appeared to partake of an order not unlike that which may be discovered in an organism. (Ghiselin 1974, p. 17)

According to Ghiselin, the old way of thinking is in the process of being completely overthrown by the evolutionary paradigm, in which all adaptations are explained in terms of their advantages to individuals. However, the Darwinian revolution is still far from complete, and the archaic teleology still creeps into the thinking of many ecologists. I agree with Ghiselin's interpretation. Consider, for instance, the following passage from Odum (1971).

In summary, although the anaerobic saprophages (both obligate and facultative) are minority components of the community, they are none-the-less important in the ecosystem because they alone can respire in the dark, oxygenless recesses of the system. By occupying this inhospitable habitat they "rescue" *energy and materials for the majority of aerobes. What would seem to be an* "inefficient" *method of respiration, then, turns out to contribute to the* "efficient" *exploitation of energy and materials in the ecosystem as a whole. (Odum 1971, p. 28).*

Similar statements claiming that the characteristics of organisms are designed to benefit the species or the ecosystem are scattered throughout the literature. Anyone who seriously considers the concept of superorganisms finds himself or herself in the peculiar situation of having many supporters, but few who have a rigorous foundation to their beliefs.

Thus, we possess a very strong philosophical bias towards seeing purpose and order in nature above the level of the individual, but there is nothing in the first principles of natural selection to justify such a bias. This may seem like a strange beginning for a book that ultimately supports the superorganism concept, but it is necessary to stress from the start that any tendency for an individual to behave in the interest of its community is something to be demonstrated, not assumed. We should not be deterred by past uncritical thinking. At the heart of the philosophical bias is an interesting scientific hypothesis that might well be reconciled with evolutionary theory. If the superorganism concept is not a first principle of natural selection, it may still be an emergent property.

The two seemingly contradictory propositions to be reconciled are: (a) an individual behaves only in its own self-interest, and (b) an individual behaves in the interest of its community. One possible solution, the one developed here, is essentially the economic argument that communities select for beneficial traits among their members. After all, an individual is surrounded by many others, each of whom is self-interested and many of whom can reward or punish that individual's activities, according to its effects on themselves. In a sense, a large part of the individual's fitness is controlled by other members of the community. This can constitute a powerful force constraining an organism to operate in the interests of the community. One need only look to human communities for a convincing example. A major portion of our activities are of direct benefit to the community but only of indirect benefit to ourselves, through the complex system of positive and negative reinforcements with which the community responds to our actions. The process is not perfect, and there are myriad ways to cheat, but these should not be overestimated. I doubt if anyone could give a satisfying description of human communities without evoking the concept of professions that are designed to be of service to others.

Cynics (e.g., Ghiselin) will reply that even if organisms are constrained to act in the interest of the community, the superorganism concept is not useful because the *real* motive is self-interest. But it is not the business of ecologists to ascribe motives to worms and algae. The statement that individuals behave to maximize their fitness may be general, but it is also empty if the actual behaviors cannot be specified. If we can show that under biologically realistic conditions, genotypes that benefit the community are selected over genotypes that do not, then we have accomplished our task. Furthermore, much more than semantics is involved. Evolutionary ecologists are forever stressing the similarity between their science and economics. But if they were really similar, then our conception of biological communities would be considerably different from what evolutionary ecologists are currently trying to describe.

For example, let us examine more closely the concept of the profession, which most would agree is fundamental to the study of economics. A profession is the major activity that the community recognizes in its individuals. It goes without saying that a person "makes his living" by his profession, and it is also obvious that in most cases a person's profession is of service to other people. That is, a doctor cures others, an architect designs buildings for others, and so on. Exceptions (e.g., jewel thieves) are eliminated upon detection. Both personal gain and service to others are implicit in the concept, and even if the first is more general, it is not necessarily the most interesting. Most interesting is the nature of the activities that are most rewarding. These are broadly synonymous with what is of greatest use to the community.

The idea of professions is out of fashion among evolutionary ecologists, but let us see what biological communities would look like if we naïvely inspected them in the same way as we do human communities. For instance, consider bark beetle communities, about which more will be said in Chapter 5. Bark beetles excavate a system of galleries under tree bark, in which they raise their larvae. The galleries soon become a habitat for an active multispecies community, consisting of nematodes, mites, other species of insects, and fungi, many of which arrive on the bodies of the bark beetles themselves. Applying the concept of professions to this community, we might predict that many species are specialized in the maintenance of the habitat and its more important occupants. Thus, we might have "janitor" species that clear away waste products or convert them into an innocuous form. We could have specialists in microclimate control, or in overcoming the tree's toxic and physical defenses. Still other species could convert resources from inedible to edible form, and others could act as guardians, keeping away or consuming intruder species that threaten the integrity of the community. We would not expect any of these activities to exist without reward for the individuals performing them, but neither could we hope to understand the nature of the activities without an appreciation of their positive effect on other species in the community. A similar analysis could be made of larger biological systems, with chains of species specialized for efficient nutrient cycling, soil retention properties, and so on.

This is an economic view of biological communities, but far from describing the evolutionary approach, it more closely approximates the ecosystem approach to ecology (e.g., Odum 1969, 1971) that the evolutionist condemns for being superorganismic in outlook! It seems as if ecology is divided into two schools, each inspecting a different side of the same coin and each claiming that its side is all that exists. The evolutionary ecologist advocates self-interest to the exclusion of function, and the ecosystem ecologist advocates function to the exclusion of self-interest. I hope that the theory developed here may help to reconcile these two schools of thought.

We now have an intuitive statement of the problem. However, in order to make real progress, it is necessary to translate the argument into mathematical form and justify the conclusions in terms of changes in gene frequency. When this is done, we encounter an irony. Although the proposed mechanism for the evolution of the superorganism relies entirely on selfish behavior, it cannot be explained by standard population genetics models and, in fact, it requires a form of group selection for its operation.

The reason why group selection is necessary can be traced to a subtle difference between the evolutionary ecologist's definition of self-interest and the intuitive meaning of the word. The evolutionary ecologist derives a definition of self-interest from the concept of *relative fitness*. Briefly stated, this means that if a trait is to be selected, it must increase the fitness of its bearer relative to the fitness of other members in the population. An absolute increase in the bearer's fitness is not enough. If the fitness of others is increased even more, then the trait will be selected against. Conversely, a trait that decreases personal fitness may be selected if it does even more damage to other members of the population. This concept is in contrast to the intuitive idea of self-interest, which places more emphasis on personal welfare without regard to its effects on others. The concept of relative fitness is a fundamental outcome of simple population genetics models that (properly) use gene-frequency change as the criterion for selection of traits, and its implications are taken seriously as meaningful statements about nature. As one influential example of its importance in evolutionary thinking, Williams (1966, p. 160) states, "Every adaptation is calculated to maximize the reproductive success of the individual, relative to other individuals, regardless of what effect this maximization has on the population."

Now let us inspect the kind of interaction by which an individual of one species controls the activities of another species to its own benefit. Consider, for instance, an earthworm that by restructuring the soil improves plant growth, which after a time lag of weeks or months pays off as increased worm food. By this interpretation, the worm is essentially farming the plants. Few laymen would term this altruistic, but if other earthworm genotypes exist in the population that enjoy the benefits of increased plant productivity without sustaining the cost of farming, they will have the highest relative fitness, and will be selected, according to standard population genetics models. Even if there is no cost to farming, the trait is at best neutral, for all share the gain. This conclusion can be generalized, for both interspecific and intraspecific interactions. The cost of controlling another member of the community is always direct, but the benefits are usually indirect, often having a substantial time lag and being vulnerable to exploitation by "freeloader" genotypes that do not share in the costs.

Several possible solutions to the problem exist. In human communities the indirect benefits are channeled back to the person that sustained the cost by an elaborate system of individual recognition. Thus, if one person pays another to do a job, he usually expects the benefits of that job to be returned to himself and not to the community at large. Unfortunately, with rare exceptions, this mechanism cannot be applied to biological communities, because most species do not possess such sophisticated abilities to associate individuals with their actions. A second solution, more relevant to the natural world, occurs when an individual is spatially isolated from its conspecifics, in which case it alone receives the indirect benefits of its activities. This is the traditional concept of mutualism, as when a damselfish feeds its anemone and thereby indirectly improves its own protection. (See Roughgarden 1975 for an excellent theoretical approach.) The mechanism behind traditional mutualism is straightforward, but because it requires that individuals be isolated from their conspecifics, it seems to be a restricted phenomenon. If the isolation were truly necessary for the evolution of these kinds of interspecific associations, we might expect them to be very important in some specific cases, but not in general. I think that this conclusion expresses the consensus of most evolutionary ecologists today.

A third possible solution involves a population genetics model that I and several others have been working on over the past few years, which I term "structured demes." This model recognizes that even when an individual interacts with conspecifics, it does so with only a very small subset of the total deme. The deme is not a homogeneous unit, but may be envisioned as being broken up into many smaller subpopulations (termed *trait groups*), in which the actual ecological interactions occur. As an example, a bark beetle may interact with conspecifics from the same gallery or the same tree, but not with those from different trees. Traditional mutualism constitutes a special case in which the trait group consists of a single individual.

As before, we use gene-frequency change within the deme as our criterion of selection, but the structuring of the deme into ecologically isolated subunits gives rise to two pathways whereby this may happen. The first operates on the relative fitness of genotypes within single trait groups. Because trait groups are by definition homogeneous units, gene-frequency change within them may be accurately approximated by standard population-genetics models. The second pathway operates on the differential productivity of trait groups. Obviously, if the trait groups vary in their genetic composition and some produce more individuals than others, then this will bias the genetic composition of the whole deme. By any conceptual standard this second pathway is a form of group selection. It exerts a powerful effect on the types of traits selected for, even when the amount of genetic variation between trait groups is low. In fact, if the theory presented here is correct, then even a random distribution of genotypes into trait groups opens the door to an economically based concept of the superorganism.

To summarize, this book attempts to establish a general mechanism whereby organisms for their own benefit evolve to enhance or inhibit other organisms in the community. It then explores the prospects for a functional interpretation of nature.

Each of the chapters is briefly described as follows. Chapter 2 explores the concept of structured demes and some of its implications. It is largely a review of the literature, but I hope that readers familiar with the subject will work through it, since some of the interpretations and conclusions will be new.

Chapter 3 analyzes the relation between group selection and altruism. In the past, these concepts have been so closely associated that disproving the routine evolution of strongly sacrificial traits appeared tantamount to disproving group selection as an important process. This reasoning was based on the observation that many group-advantageous traits necessarily involve a sacrifice on the part of some group members. I do not dispute this observation, but question whether the sacrifice must fall on the individuals controlling the behavior. As an example, decreasing group size can be accomplished by voluntary emigration of certain individuals or by the expulsion of some individuals by others. Both solutions have identical effects on group size and both involve an identical sacrifice by group members (assuming that dispersal is individually disadvantageous). However, in one case the sacrifice falls on the individuals initiating the behavior, which makes it strongly altruistic, while in the second case the sacrifice is imposed on other individuals by the initiators of the behavior, which makes that behavior—superficially at least—selfish. Actually, as I attempt to show in Chapter 3, these behaviors can more profitably be understood as being genetically weakly altruistic, requiring a form of group selection for the same reason as given above for the evolution of interspecific control. In any case the theory presented here acknowledges that strongly altruistic traits will rarely be observed in nature, but still predicts that group selection can be a powerful evolutionary force by operating through interference behavior. In some cases natural selection may actually maximize trait-group productivity. Chapter 3 dwells mostly on the group selectionist's traditional question—population regulation—with brief excursions into sociobiology and the evolution of (ecologically) complex life cycles.

Chapters 4 through 6 extend group selection theory to interspecific relationships, and explore the implications for models of multispecies coevolution. Chapter 4 develops the economic argument and sets it into a mathematical form by drawing the distinction between the direct and indirect consequences of traits. Mathematically, the direct effects appear in the equations governing the fitness of the organisms that possess them (the rows of the community matrix), while the indirect effects appear in the equations governing the fitness of other species in the community (the columns of the community matrix). Traditional models inspect only the evolutionary forces that occur within a single trait group, and conclude that natural selection is insensitive to the evolution of indirect effects. When the effects of structured demes are included, these same models predict that indirect effects are selected whenever they are advantageous for the individual performing them, regardless of the individual's relative fitness in its trait group.

In my opinion, this simple modification of existing theory has profound implications for the way we look at biological communities. The evolution of interspecific relationships based only on direct effects is severely limited—confined largely to competition and predation. However, if natural selection is sensitive to indirect effects, then organisms may routinely evolve to modify the community *through any pathway* that increases their fitness. Structured deme theory, therefore, predicts a far greater diversity of adaptive relationships among species than is currently appreciated. Chapter 5 examines some of these relationships and the prospects for a functional approach to communities. Much of the chapter is built around a "real world" example, the bark beetle community mentioned earlier. Chapter 5 concludes with a brief discussion of the concept of specialization and some shamelessly premature speculation on a possible application to biological control.

Chapter 6 returns to a purely theoretical plane. With a series of simulation models, I explore the tendency of communities whose members are motivated entirely by self-interest (in the intuitive sense of the word) to evolve into mutualistic networks. In the narrow context of the models, mutualism as an emergent property of selfish behavior is a surprisingly robust phenomenon, but the models are still a poor caricature of the real world.

Some Cautions

The group selection controversy is now over 15 years old, and the way in which it is discussed has become rather canalized. Since the theory presented here is different from the traditional paradigm, both in its mechanism and in its predictions, I would like to emphasize that many of the arguments applied against group selection in the past are not automatically pertinent to this theory.

My first caution pertains to the use of biological examples to support the models. Normally, one who constructs a theory is interested in two things: (a) the set of data that can be interpreted by the theory, even if it can also be interpreted by other theories. These data give an idea of the theory's scope and also allow one tentatively to explore implications; and (b) the set of data that can be interpreted only by the theory. These data give an indication of the theory's generality relative to competing theories.

Obviously, the latter kind of data is of greater interest, but it is often difficult to obtain, and in its absence one is forced to evaluate carefully the relative plausibility of competing hypotheses. Because of the lack of precise data, most ecological and evolutionary theorists find themselves in this situation, and those theorists studying group selection have been no exception. G. C. Williams carefully evaluated the group selection controversy in 1966 and concluded that the mechanism for group selection proposed at that time was so weak that an argument for individual selection should be preferred whenever it can reasonably explain the observations. Unfortunately, Williams' opinion, though fully justified, has degenerated into a pointless game whereby any conceivable individual advantage of a behavior becomes the chosen interpretation, regardless of its plausibility.

Since the mechanism proposed here is fundamentally different from traditional group-selection theories, I hope both it and the examples used to illustrate it will be given the careful consideration accorded to most ecological and evolutionary models. To facilitate this, I stress that most of the examples cited do have alternative interpretations. I intend these examples as illustrations of how the process might operate in nature, and not as evidence for the process. However, this does not mean that the theory presented here is untestable. In fact, one example (the insect-mite interactions studied by J. C. Moser and co-workers, described on page 119) is sufficiently precise and pronounced in its pattern that I consider it as solid support, most parsimoniously explained by the models developed in this book.

My second caution concerns the relation between levels of natural selection. Group selection and individual selection are often looked upon as opposing forces, no doubt because of the emphasis placed on self-sacrificial traits. The conflict between levels of selection is much less apparent in the models presented in this book. For natural selection to be sensitive to group productivity, the deme must be divided into ecologically isolated trait groups, but given that sensitivity, any conflict between the group and individual components of natural selection will be minimized whenever possible. As a result, organisms appear to enhance individual and group fitness simultaneously. I look upon the levels of selection as levels of organization, rather than as opposing forces. Selfish genes organize themselves into selfish individuals, which organize themselves into selfish populations, which organize themselves into selfish multispecies communities. This process is described in more detail in Chapter 6.

Finally, a word should be said about the readers for whom this book is intended. I am a firm believer in the value of all approaches to ecology, from mathematical modeling to basic field studies. The intuition of field ecologists must be translated into precise mathematical language, made consistent with the existing laws of evolutionary theory, and then tested in the laboratory and field. Each phase has grown sophisticated over the last decade, and I have unfortunately been unable to master the full spectrum. Perhaps foolishly, I have still tried to communicate with the full spectrum in this book. To me, it seemed as if I had no choice, since I know what each phase has to offer in the refinement of the ideas. No doubt, specialists will complain about my naïvete in their own areas of expertise. However, if I can show that a theory of superorganisms has enough substance to make it worthwhile for others to improve upon its deficiencies, I will have accomplished my goal.

TABLE 1.1 Terms used in book

Chapter 2		
A, B	Two "types" (alleles in a haploid model) that constitute the population	
N	Total density of the population	
p, q	Frequency of A-types and B-types, respectively $(p + q = 1)$	For models assuming constant trait group density
P_p	Proportion of trait groups containing a frequency of p A-types	
$\sigma^2{}_p$	Variance in the frequency of A types among trait groups	
p_A, p_B	Average subjective frequencies: the frequency of A experienced by the average A-type and B-type in the global population	
m, n	Number of A- and B-types, respectively, in a single trait group	For models assuming variable trait group density
$P_{m,n}$	Proportion of trait groups containing m A-types and n B-types	
σ^2_m	Variance in the number of A-types among trait groups	
m_A, m_B	Average subjective densities: the number of A experienced by the average A- and B-type in the global population	

TABLE 1.1 (Continued)

f_A, f_B	Absolute individual fitness of *A*- and *B*-types within a single local population
F_A, F_B	Average absolute fitness of *A* and *B*-types for global population
T	Number of trait groups in the global population
d	Effect of an *A*-type on itself
r	Effect of an *A*-type on every other member of the local population
Chapter 3	
Z	Population size at which the trait group crashes to extinction
w	Constant governing the maximum productivity of the trait group
t	Time
N_G, N_J	Density of adults and juveniles (larva), respectively
w_G, w_J	Feeding rate of adults and juveniles, respectively
k_1	Proportion of *A*-types remaining in the trait group after voluntary removal
k_2	Ratio of individual *A*-type's fecundity/individual *B*-type's fecundity
k_3	Per capita inhibition of *A*-type on entire trait group (including itself)
k_4	Per capita inhibition of *A*-type on other *A*-types
lk_4	Per capita inhibition of *A*-type on *B*-types
k_5	Per capita stimulation of *A*-type on entire trait group
Chapter 4	
s	Number of species in the community
a_{ij}	Per capita effect of species *j* on species *i*
c	Cost of an *A*-type's activity to itself
g	Gain of an *A*-type's activity to itself, plus to every other member of the trait group
N_x	Density of species x
M_x	Rate of increase of species x
K_x	Carrying capacity of species x
L	Constant governing the rate at which carrying capacities asymptote
I	Amount of time spent within trait group
VN_x	Variance in density of species x among trait groups
Q	Number of trait group "generations"
Chapter 5	
R_i	Abundance of resource *i*
U_i	Resource abundance that would exist if the population consisted entirely of species *i*
p_i	Proportion of species *i* in the community
Chapter 6	
u	Range of values for matrix elements
z	Constant governing rate of evolution
w	Adaptive potential (capacity to effect fitness of community)

(continued)

TABLE 1.1 (Continued)

v	Constant to determine direction of evolution
A	Matrix of interaction coefficients
a_{ij}	Per capita effect of species j on species i
E	Matrix of newly evolved effects
e_{ij}	Newly evolved effect of species j on species i
B	Species-welfare matrix
b_{ij}	The effect of incrementing the corresponding element a_{ij} on the density of species j
C	Community-welfare matrix
c_{ij}	Effect of incrementing the corresponding element a_{ij} on the density of the entire community

2 Individual Selection and the Concept of Structured Demes

The Basic Selection Model

Most people are familiar with the basic theory of natural selection. Organisms vary in a heritable fashion. Some variants leave more offspring than others; their characteristics, therefore, are represented at a greater frequency in the next generation.

If one were allowed the luxury of renaming theories, a good name for this one might be the "principle of individual selection." But individual selection as we know it is not equivalent to natural selection. The modern concepts of individual selection, group selection, and all their problems arise at the next link in the chain of reasoning, with the questions, "What are those traits that allow the individuals possessing them to leave more offspring than others? What constitutes fitness in nature?"

To answer these questions, evolutionary biologists built a very simple model, whose properties are well described by Mayr (1963): "Under the impact of modern systematics and population genetics, a usage is spreading in biology that restricts the term 'population' to the *local population*, the community of potentially interbreeding individuals in a given locality. All members of a local population share in a single gene pool, and such a population may be defined also as a 'group of individuals so situated that any two of them have equal probability of mating with each other and producing offspring.'"

In this manner, the bulk of Darwinian theory has been built around the evolutionary forces that operate within local populations, or demes. The mathematical expression of this concept follows an unvarying format. Two types, A and B (these may be alleles, haploid organisms, or genotypes) exist at a combined density of N and in frequencies p and q ($p + q = 1$), respectively. The types are assigned fitnesses (f_A and f_B) which can be constants or functions with any number of variables. By definition, the type with the highest fitness increases its proportionality in the next generation.

Hereafter, the preceding formulation will be referred to as the *basic selection model.* It sounds like an exact mathematical statement of evolution's fundamental rules, and it is—but with at least one additional assumption. By specifying a single frequency of types in the population, the basic selection model expressly forbids the situation in which some individuals experience a frequency of p_1 and others a frequency of $p_2 \neq p_1$. In other words, it assumes a spatial homogeneity in the genetic composition of populations.

A large literature is devoted to relaxing the homogeneity assumption in ecology and population genetics, but the basic selection model still enjoys a special prominence in evolutionary theory, and for good reason. It should be remembered that Darwin and his first followers were devoted to overthrowing a completely different paradigm—special creation. To them, the answer to "what constitutes fitness?" seemed so obvious in so many different ways that they hardly needed to build a model at all. Clearly a bat with wings, a predator with sharper eyes, a plant with a more efficient root system, or a giraffe with a long neck would be more successful in its struggle against nature than others without such endowments. In other words, the first evolutionary biologists were preoccupied with structures and behaviors that are a constant advantage to the individuals who possess them. These traits can be modeled by using constants for f_A and f_B. With constant fitness terms, the expected type is always favored in the basic selection model. A model is only as good as its predictions, and for constant fitness terms, the basic selection model provides a very exact fit to what seems to happen in nature. Of course, one could build many other models that do not assume spatial homogeneity in genetic composition and that also produce the same conclusions, but the basic selection model is the simplest. Why add complexity when it doesn't increase predictive value?

In other words, the homogeneity assumption is not *necessary* for the selection of traits with constant fitness values; it is simply the easiest way to express their selection. This is a critical point. In fact, the basic selection model can now be resummarized as follows: It consists of evolution's fundamental rules plus an additional assumption of spatial homogeneity. The homogeneity assumption is neither necessary nor realistic. It is simply made because it is convenient and because it doesn't seem to interfere with the selection of traits that can be represented by constant fitness values.

As mentioned previously, the effects of relaxing the homogeneity assumption have been carefully and systematically explored for many areas of evolutionary theory, but not for all of them. Gaps exist in which homogeneity is assumed without the effects of relaxing it being fully known. As I will argue in this book, the concept of individual selection, on which many biologists base their understanding of social behavior and interspecific interactions, represents one such area.

Today the basic fact of evolution is taken for granted, and interest has shifted to more subtle classes of adaptations. One of these classes concerns traits that influence not only the fitness of the individuals possessing them, but other members of the population as well. These traits include overtly social behaviors, such as aggression or cooperation, and also activities that through some modification of the biotic and abiotic environment feed back to affect the population at large (e.g., pollution, resource depletion). If we let d represent an individual A-type's effect on itself and define r as its effect on every other member of the population, then changes in fitness due to the A-type's trait can be calculated in a simple linear haploid basic selection model:

$$\begin{aligned} f_A &= d + (Np - 1)r \\ f_B &= Npr \end{aligned} \qquad (2.1)$$

(recall that N is the total density of both types and p is the frequency of the A-type). The term f_A consists of an A-type's effect on its own fitness plus the fitness it receives from $(Np - 1)$ other A-types in the population. The B-type has no effect on itself and serves as a recipient for Np A-types. If the types are equal in other respects, then the A-type is selected only when $f_A > f_B$. That is, when

$$d + (Np - 1)r > Npr \qquad (2.2)$$

$$d > r \qquad (2.3)$$

To be favored by selection, an individual's effect on itself must be more positive than its effects on others. Although superficially reasonable, this conclusion has some disturbing implications. It predicts that on the most fundamental level there is nothing preventing the following events from routinely occurring in nature:

1. If an individual decreases its own fitness (negative *d*), it can be selected for, as long as it decreases the fitness of others even more. (This phenomenon has been called "spite"; see Verner 1977 for a recent hypothesis concerning spite in the evolution of territoriality.)
2. If an individual increases its own fitness (positive *d*) it will be selected against if it increases the fitness of others even more.

In short, this basic selection model predicts that natural selection is insensitive to the fitness of the population as a whole. *Natural selection is sensitive only to the fitness of a genotype relative to other genotypes within that population.* This feature is such a fundamental part of basic selection models in general that the majority actually define fitnesses relative to the most fit type in the population, thereby eliminating a consideration of population productivity from the start.

The concept of *individual selection* is not precise enough to define rigorously (see Williams 1966, Alexander 1974, E. O. Wilson 1975 for detailed discussions), but I think that it is accurately represented by this model and its properties, especially if kin selection is temporarily ignored. Throughout this book the term individual selection will refer to the above criterion, or stated in words: "To be selected, a type must have the highest fitness, relative to other types in the local population." The word "local" is central to the definition, as will become clear below.

The consequences of individual selection disturb some evolutionists because they mean, among other things, that even while increasing their relative fitness, the individuals in a population could be evolving to extinction (Hamilton 1971, Roughgarden 1976). I will argue later that there is an even more important drawback, in that enormous pathways toward the evolution of increased fitness are blocked. Neither of these predictions correspond to the biologist's intuitive idea of evolutionary adaptation.

One would think that the conflict could be resolved by going out in nature and seeing what the organisms are actually doing, but this is the single most discouraging thing about the group-selection controversy. Field studies and observations are almost never precise enough to provide measurements of fitness functions, especially for subtle adaptations that affect the population at large. Laboratory studies may be precise, but unfortunately they are self-fulfilling prophecies, designed to mimic either the homogeneity assumption of basic selection models or the special assumptions of its alternatives. Moreover, even if it were shown in a single or a few cases that natural selection promotes the fitness of the whole population, the individual selectionist can explain the results by saying that many traits which increase relative fitness coincidentally increase the fitness of the population at large; therefore, even though natural selection is insensitive to the latter, it may still be occasionally observed on the basis of the correlation (e.g., Ayala 1968). This is certainly true of the obviously advantageous traits modeled by constant fitness terms, but is more questionable for the traits represented by equation 2.1. I do not mean to imply that methods of field falsification are impossible (some are suggested later). The real problem is that without a coherent alternative model, no one knows where to look for solutions.

Again, we can summarize by making the following statements about the basic selection model. It makes predictions about a large class of traits (constant fitness terms) that conform to most people's intuition. This accounts for the prominence of the model in evolutionary theory. However, it also makes predictions about another large class of traits (in the form of the individual-selection model), over which there is much controversy.

Clearly, there are only two reasonable solutions to the problem. Either individual selection's predictions about relative fitness are correct, or they are an artifact of the homogeneity assumption.

Relaxing the Assumption

The key to group selection, if it exists, is to be found by relaxing the homogeneity assumption. But how? Population geneticists have developed several conceptions of spatial heterogeneity for a variety of purposes, such as for the role of genetic drift in gene-frequency variation, the effect of habitat heterogeneity on protected polymorphisms, and the nature of geographic clines. (See Felsenstein 1976 for a review of these theories.) The three most prevalent models of spatial heterogeneity may be characterized as follows:

1. *Island models* consider a global population composed of a number of spatially isolated local populations, each of which is internally homogeneous and described by the basic selection model. However, separate local populations may differ in any parameter, including the frequency of genotypes. The local populations are connected to each other by a flow of dispersers. A given proportion of each population migrates, and each disperser moves to a new population at random, disregarding geographical distances. The island model forms the foundation for traditional group-selection theories. Wright (1938, 1945) originally applied the model to the evolution of altruistic traits and thought the island model to be "the

greatest creative factor of all" in evolution. The idea was made controversial by Wynne-Edwards (1962) and made sophisticated by several other authors (e.g., Levins 1970, Eshel 1972, Boorman and Levitt 1973, Levin and Kilmer 1974, Gilpin 1975). See E. O. Wilson (1973; 1975, ch. 4), Maynard Smith (1976), and Wade (1978a) for reviews.

2. One modification of the island model is to limit the dispersal of individuals to neighboring populations, in which case a favorable mutation slowly spreads outward from its original local population. This is usually termed the *stepping-stone model*, and mainly has been used to explore the spread of favorable mutations (with constant fitness values) and the existence of clines (Felsenstein 1976). Continuous state approximations of stepping-stone models also exist in which individuals are not grouped into discrete local populations. Neither have been applied to the group-selection controversy, although Slatkin and Wade (1978) have relaxed the island model's restrictive assumptions about random migration in another way.

3. A limiting case of the island model exists when dispersal is total; in other words, when all individuals periodically leave their local population and disperse throughout the global population. This is usually termed the *subdivided population model*. Although it is formally a special case of the island model, it is important to stress that biologically it represents a substantially different conception of population structure. The island model with low dispersal describes populations that are permanently separated in space. Dispersal between local populations is thus an *interdemic* process. In subdivided population models, the local populations are more ephemeral. In a sense, the dispersal of all individuals throughout the global population satisfies the definition of a single deme, in which case the subdivision into local populations may be regarded as an *intrademic* process. More will be said about this distinction later.

The subdivided population model was first considered by Wahlund (1924) to study its effects on the frequency of genotypes in a one-locus, two-allele system without selection. To date, its most important application was initiated by Levene (1953) who explored the effect on the maintenance of genetic polymorphism in a heterogeneous environment. (See Christiansen and Feldman 1975 for a modern review). Two habitat types are specified (or more generally "niches"), each of which bestows a selective advantage upon different alleles of a one-locus, two-allele system. After random mating, zygotes are cast over the global population, after which selection occurs within each local habitat. Two types of selection are often considered: so-called "soft" selection, in which each local habitat produces the same number of individuals; and "hard" selection, in which the productivity of local habitats may differ. Notice that the central focus of this application is spatial heterogeneity in the *habitat* and its effect on the maintenance of polymorphisms. This is quite different from the models described below, which focus on spatial heterogeneity in the genetic composition of populations in a uniform habitat.

In the context of the group-selection controversy, subdivided populations have been an implied assumption of kin selection and inclusive fitness models (Hamilton 1964, 1971, 1975) and have been explicitly considered by Charnov and Krebs (1975), Cohen and Eshel (1976), Eshel (1977), Matessi and Jayakar (1973, 1976), and Wilson (1975a, 1977a). These models form the mechanism behind the theory presented in the book, and will be described in more detail in the following paragraphs.

Before we explore the effect of any specific form of population structure on the individual-selection model, it should be stressed that all forms have the same qualitative effect. In all cases, the change in gene frequency within a local population depends upon the evolutionary forces operating within that population (described by the basic selection model) and on some contribution from the outside in terms of dispersers from other local populations. The gene frequency among dispersers is determined by two factors: (a) the gene frequency of the local populations from which they are derived, and (b) the density of the local populations from which they are derived (i.e., a dense population will send forth more dispersers than a sparse one). Consider a genotype whose activities increase the productivity of its local population without, however, changing the gene frequency within the population (hard selection, by definition). Populations with a high frequency of this genotype will be more productive than those with a low frequency, and will differentially contribute to the pool of dispersers. The genetic composition of the dispersers will be biased toward the genotype that increases the productivity of its group, and this bias is carried into all groups colonized by the dispersers.

To summarize, when spatial heterogeneity is incorporated into population-genetics models, natural selection becomes sensitive not only to the productivity (fitness) of individuals relative to others in their local population, but also to the productivity of local populations relative to others in the global population. This latter component may be regarded as natural selection on the level of populations, or *group selection*. The term "group selection" is usually applied to the more specific island model of Wright as interpreted by Wynne-Edwards (Maynard Smith 1976), but it accurately describes the other models as well (Wade 1978b, Wilson 1979a).

Since the differential productivity of groups is an inevitable consequence of relaxing the homogeneity assumption, the central question of group selection is not whether it exists, but how strongly it operates, a thesis most authors have recognized (e.g., Williams 1966, E. O. Wilson 1975, Maynard Smith 1976). For instance, in traditional group-selection models, dispersal between local populations is relatively low. Any change in gene frequency caused by a disperser colonizing an established population is negligible, and a significant impact requires the periodic extinction of populations with colonization by a few (but not too many) dispersers. The balance of parameter values necessary to produce a significant effect of group selection on gene frequency is sufficiently delicate that so far traditional models have been rejected as a powerful evolutionary force.

The Concept of Structured Demes

It is beyond the scope of this chapter to evaluate the effect of all forms of population structure on the individual-selection model. Instead, my goal is to identify one form of population structure that is fairly general (in other words, applies to many if not most species in nature) and to explore the implications of this structure for the individual-selection model. To do this, it is necessary to examine the deme concept closely. What exactly does it imply?

The deme is defined as a population that is homogeneous with respect to the mixing of genes, but with frequency-dependent fitness values, the basic selection model assumes more than this definition. Actual selection models say nothing about the mixing of genes, but refer instead to the manifestation of traits. In particular, models in the form of equation (2.1) refer to the *ecological effect* of A-types on themselves and other members of the population. Therefore, in addition to creating a localized evolutionary arena with equal mating probabilities, basic selection models also specify that the population is homogeneous with respect to the manifestation of traits. They assume that each and every individual in a deme experiences, and thereby has its fitness determined by, exactly the same frequency of genotypes.

Is there any such thing as a population that is homogeneous with respect to the manifestation of traits? Yes, but it is usually rather small. Perhaps this point can be made clear with a series of examples:

1. The pitcher plant mosquito *Wyeomyia smithii* lays its eggs in the water-holding leaves of the pitcher plant. Each pitcher is an isolated habitat unit, usually containing from 1 to 50 mosquito larvae and from 4 to 50 ml of water. Almost certainly the pitcher represents a homogeneous population with respect to the manifestation of larval traits. Each individual feels the effect of every other individual. If any aspect of the social, biotic, or abiotic environment (such as resource depletion, toxin production) is a function of the genotypic composition of the population, all individuals will experience its effects equally. However, there is no reason to expect larval populations in different pitchers to have the same genotypic composition. They will certainly differ, if only at random.
2. A bark insect finds a vulnerable tree and emits a pheromone that attracts conspecifics to the scene. The pheromone can only be detected within a certain radius of its source. All individuals within this radius benefit equally. No one outside the radius benefits. Similarly, once the galleries have been formed, the community within a single gallery may be homogeneous in terms of the effects of genotypes, but will certainly differ from the genotypic composition of other galleries in other trees.
3. A school of fish may be homogeneous for certain traits, yet many schools coalesce to shoal.
4. A female bird has offspring which for a time share a nest. Undoubtedly every sibling interacts with the same individuals in the nest as does every other sibling, but this interaction has no bearing on the interactions within other nests.
5. Many marine invertebrates have pelagic larval stages which later settle to complete development. The deme for such species is very large, sometimes constituting thousands of square miles. However, ecological interactions take place within much smaller populations, consisting of an individual and a few of its neighbors. The boundaries around these smaller populations may not be discrete, as they are for examples 1–4, but lack of boundaries does not appear to be important, as will be demonstrated below.

In sum, every trait has a "sphere of influence" within which the homogeneity assumption is roughly satisfied. It is the area within which every individual feels the effects of every other individual. I have termed the population within this area a *trait group* (Wilson 1975a, 1977a) to emphasize its dependence on the particular traits being manifested. For instance, the trait group for traits manifested among nestling birds consists of those individuals within a single nest, but the trait group for traits manifested during adulthood consists of a much larger population.

The individual-selection model with its assumption of spatial homogeneity in genetic composition is a viable, realistic concept when applied to single trait groups. But in all the above examples, the trait groups are very small—far smaller than the deme, or population within which genetic mixing occurs. The pitcher plant mosquito larvae metamorphoses into an adult and flies throughout the bog. Fish shoal. Marine invertebrate larvae are pelagic. Birds leave their nests. Insects travel long distances on dispersal flights.

A moment's consideration should convince the reader that this fact is true in general. Trait groups are almost always smaller than the deme, and for a very general reason. Most organisms tend to concentrate their movement in a brief dispersal stage—the seeds and pollen of plants, the post-teneral migrating stage of insects (Johnson 1969), the larvae of benthic marine fauna, the adolescents of many vertebrates. At all other periods of the life cycle, the individuals are relatively sedentary, their movements trivial in scale compared to their dispersal stage. Interactions in the nondispersal stages will be within trait groups smaller than the deme.

The relation between the trait group and the deme also holds for organisms that do not have well-defined dispersal stages. Consider nonmotile microorganisms, such as some soil bacteria that rely on passive dispersal. For a period of generations, single populations will develop on microsites for which the homogeneity assumption is satisfied. However, periodic physical disturbances, such as rain storms, will result in a massive mixing of local populations. Alternatively, consider a population of organisms that mix continuously throughout their life cycle, such as zooplankton. Any trait manifested by an individual that, by the nature of the trait, influences only a small number of surrounding individuals, will necessarily substructure the deme into trait groups, even if the membership of the trait groups is constantly changing.

To summarize, it may be said that while the individual-selection concept is valid, it is valid only for tiny populations (trait groups). Any realistic evolutionary model in which fitness is frequency-dependent must recognize two population units: one that is homogeneous with respect to ecological interactions (the trait group), and another that is homogeneous with respect to genetic mixing (the deme). Even a deme is composed not only of a population of individuals, *but also of a population of trait groups.* The trait groups are isolated with respect to the manifestation of traits, but periodically mix and resegregate due to the dispersal process. While isolated, the trait groups can vary in their genetic composition.

The distinction between the trait group and the deme is probably obvious to most biologists, and has also been recognized in many theoretical models (e.g., Trivers 1971, p. 44), but it has not been sufficiently emphasized as a biological generality. In particular, this distinction is ignored by any mathematical model (and the thinking behind it) that assigns a single frequency to its genotypes, e.g., the basic selection model. In my opinion, the concept of trait groups represents a biologically realistic form of population structuring that describes many if not most species in nature (Wilson 1977a, 1979). It is best described by the subdivided population model with hard selection, and to differentiate it from other uses of this model, I have termed it a theory of "structured demes" (Wilson 1975a, 1977a). Interdemic forms of population structure with low migration, such as the island model, may of course also occur. But in the context of individual selection, each island should be recognized as being further differentiated into trait groups.

Let us explore the effects of deme structure on the predictions of individual selection: first, in terms of the simple linear fitness functions developed on page 15, and then more generally. In equation (2.1), the individual fitness of the A and B types was represented as

$$\begin{aligned} f_A &= d + (Np - 1)r \\ f_B &= Npr \end{aligned} \qquad \textbf{(2.1)}$$

where d is the effect of an individual A-type on itself and r is its effect on every other member of the population. These two equations may now be taken to represent the fitnesses within a single trait group.

In structured demes we have a large number of trait groups (T), each of which can be characterized by a single density and frequency. Assume that T is large and let P_{mn} = the proportion of trait groups containing m A-types and n B-types ($m + n = N$). The fitness of the two types over the entire deme, designated F_A and F_B, consists of the weighted average of the fitness over all the trait groups:

$$\frac{T \sum_{m,n=0}^{\infty} P_{mn} m \left[d + (m - 1)r\right]}{T \sum_{m,n=0}^{\infty} P_{mn} m} \tag{2.4}$$

$$F_B = \frac{T \sum_{m,n=0}^{\infty} P_{mn} nmr}{T \sum_{m,n=0}^{\infty} P_{mn} n} \tag{2.5}$$

To average over the trait groups is biologically justified because of the physical mixing of the organisms during the dispersal stage. As before, the *A*-type is selected only when it has a higher relative fitness over the *global* population (the deme) than the *B*-type, although it need not necessarily have a greater fitness within the local population (the trait group), as we shall see.

$$F_A > F_B$$

$$d + \frac{\sum_{m,n=0}^{\infty} P_{mn} m^2 r}{\sum_{m,n=0}^{\infty} P_{mn} m} - r > \frac{\sum_{m,n=0}^{\infty} P_{mn} nmr}{\sum_{m,n=0}^{\infty} P_{mn} n} \tag{2.6}$$

Rearranging:

$$d > r \left(\frac{\sum_{m,n=0}^{\infty} P_{mn} nm}{\sum_{m,n=0}^{\infty} P_{mn} n} - \frac{\sum_{m,n=0}^{\infty} P_{mn} m^2}{\sum_{m,n=0}^{\infty} P_{mn} m} + 1 \right) \tag{2.7}$$

$$d > r \left[\frac{E(mn)}{E(n)} - \frac{E(m^2)}{E(m)} + 1 \right] \tag{2.8}$$

Let $m = E(m), \bar{n} = E(n)$

$$\sigma_m^2 = E\left[(m - \bar{m})^2\right], \sigma^2 = E\left[(n - \bar{n})^2\right]$$

$$\sigma_{mn}^2 = E\left[(m - \bar{m})(n - \bar{n})\right]$$

Then

$$E(mn) = \sigma_{mn}^2 + \bar{m}\bar{n}$$

$$E(m^2) = \sigma_m^2 + \bar{m}^2$$

and (2.8) becomes

$$d > r\left(\frac{\sigma^2_{mn} + \bar{m}\bar{n}}{\bar{n}} - \frac{\sigma^2_m + \bar{m}^2}{\bar{m}} + 1\right) \tag{2.9}$$

$$d > r\left(\frac{\sigma^2_{mn}}{\bar{n}} - \frac{\sigma^2_m}{\bar{m}} + 1\right) \tag{2.10}$$

Assume that the initial density in each trait group is constant (i.e., $m + n = N$). In that case, $\sigma^2_{mn} = -\sigma^2_m$ and (2.10) becomes

$$d > r\left[-N\left(\frac{\sigma^2_p}{\bar{q}} + \frac{\sigma^2_p}{\bar{p}}\right) + 1\right] \tag{2.11}$$

where σ^2_p is now the variance in relative frequency of the *A*-type. Alternatively, let trait-group density vary, but the covariance between *m* and *n* equals zero ($\sigma^2_{mn} = 0$). Then

$$d > r\left(1 - \frac{\sigma^2_m}{\bar{m}}\right) \tag{2.12}$$

Notice that equations (2.11) and (2.12) resemble the criteria for selection of the *A*-type in the individual-selection model [equation (2.3)], with the addition of a term that contains the variance in the frequency of the *A*-type between trait groups. This is the "between trait group" component of natural selection, whose significance will be explored in detail below.

The same result may be achieved through a slightly different pathway, by calculating the frequencies experienced by the average *A*- and *B*-type in the deme. These may be termed the "average subjective frequencies" (p_A, p_B: Wilson 1977a) where

p_A = the frequency of *A* experienced by the average *A*-type
p_B = the frequency of *A* experienced by the average *B*-type

If we assume a constant starting trait-group density, and let P_p represent the proportion of trait groups with a relative frequency of p, then

$$p_A = \frac{\sum_{p=0}^{1} P_p p^2}{\sum_{p=0}^{1} P_p p} = \bar{p} + \frac{\sigma^2_p}{\bar{p}} \tag{2.13}$$

$$p_B = \frac{\sum_{p=0}^{1} P_p pq}{\sum_{p=0}^{1} P_p q} = \overline{p} - \frac{\sigma_p^2}{\overline{q}} \tag{2.14}$$

The fitness of average individuals are

$$\begin{aligned} F_A &= d + (Np_A - 1)r = d + \left[N(p + \frac{\sigma_p^2}{\overline{p}}) - 1\right]r \\ F_B &= Np_B r = N(p - \frac{\sigma_p^2}{\overline{q}})r \end{aligned} \tag{2.15}$$

and the A-type is favored by selection when

$$F_A > F_B$$

which can be shown to equal

$$d > r\left[-N(\frac{\sigma_p^2}{\overline{q}} + \frac{\sigma_p^2}{\overline{p}}) + 1\right] \tag{2.16}$$

which is identical to equation (2.11). An "average subjective density" may also be calculated as $\overline{m} + \sigma_m^2/\overline{m}$, which is identical to Lloyd's (1967) concept of "mean crowding."

Average subjective frequencies are useful because they allow one to consider the effects of deme structure in any linear basic selection model simply by replacing the average frequency ($\overline{p}$) with the average subjective frequencies (p_A, p_B) without going through calculations (2.6–2.12) in every case. Average subjective frequencies also supply a biologically understandable way to view the effect of structured demes on the natural selection of traits. This is best illustrated by a numerical example. Figure 2.1 shows a deme subdivided into four trait groups, each containing five individuals. The frequency of A varies among trait groups. The average frequency of A over the whole deme may be calculated as

FIGURE 2.1 Four discrete trait groups that vary in their composition of A- and B-types. Ecological interactions occur within trait groups, but individuals from all trait groups mix during a dispersal stage.

$$\bar{p} = \frac{0.2 + 0.4 + 0.6 + 0.8}{4} = 0.5$$

However, the average subjective frequencies of the A- and B-types are

$$p_A = \frac{1(0.2) + 2(0.4) + 3(0.6) + 4(0.8)}{10} = 0.6$$

$$p_B = \frac{4(0.2) + 3(0.4) + 2(0.6) + 1(0.8)}{10} = 0.4$$

The subjective frequencies, therefore, are simply the weighted averages over all the trait groups. In other words, only 1 A-type experiences the low frequency of 0.2 A-types (itself), but 4 A-types experience the high frequency of 0.8 A-types, and so on. It may easily be verified from equations (2.13) and (2.14) that any variance in frequency among trait groups causes the average A-type to experience a higher frequency of A-types than actually exists in the deme, while the average B-type experiences a lower frequency (i.e., $p_A > \bar{p} > p_B$ and, conversely, $q_B > \bar{q} > q_A$). Fisher (1958), Hamilton (1963, 1964, 1970, 1971, 1975) and Trivers (1971) realized that altruistic traits which decrease the fitness of the donor to the benefit of the recipients can be favored by selection only if they are directed toward fellow altruists—in other words, toward similar "types." The average subjective frequencies show that to some extent this occurs in all structured demes, and that the strength of the process depends upon the variance in frequency among trait groups (hereafter referred to as trait-group variation).

More specifically, Figure 2.2 represents a continuum of trait-group variation. Several points on this continuum are of special interest:

Point 1. When no trait-group variation exists, equation (2.16) reduces to $d > r$, the individual-selection model. Mathematically, of course, $\sigma_p^2 = 0$ is nothing more than a return to the homogeneity assumption, but biologically, it is important to recognize that if one accepts the basic concept of structured demes, individual selection represents an absurd extreme of the continuum. Variance within trait groups is zero by definition, but zero variance between trait groups would require a very special mechanism indeed.

Point 2. In the absence of good information, the most conservative assumption that can be made about trait-group variation is that it is random, like tossing red and blue marbles into urns. In this case, the variance can be approximated by the binomial distribution ($\sigma_p^2 = \bar{p}\bar{q}/N$). Substitution of this value into (2.16) becomes

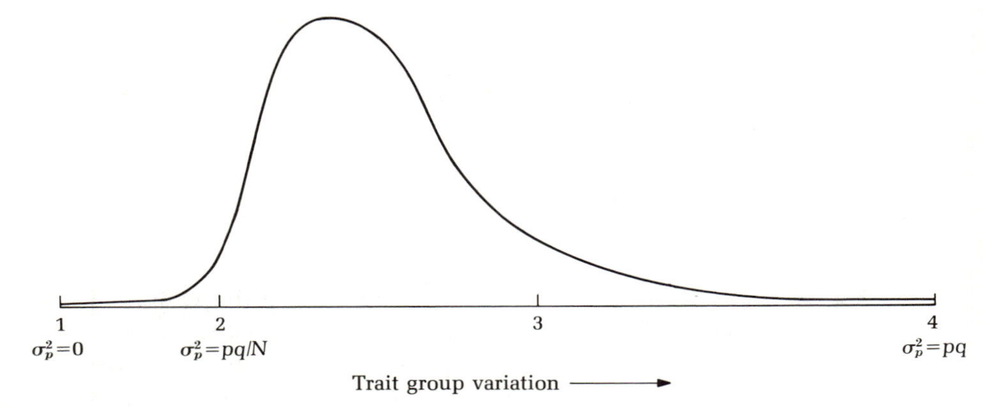

FIGURE 2.2 A continuum of trait group variation (variance in the frequency of *A* among trait groups). Points 1 through 4 correspond respectively to individual selection, the binomial distribution, kin selection, and total altruism. The curve gives a hypothetical distribution of populations in nature along this gradient.

$$d > r\left[-N\left(\frac{\overline{p}\,\overline{q}}{N\overline{q}} + \frac{\overline{p}\,\overline{q}}{N\overline{p}}\right) + 1\right] \tag{2.17}$$

$$d > r\left[-(\overline{p} + \overline{q}) + 1\right]$$

but $\overline{p} + \overline{q} = 1$, so

$$d > 0 \tag{2.18}$$

With variable density and $\sigma_m^2 = \overline{m}$, the same result may also be shown using equation (2.12). Stated in words: Given a binomial distribution of types into trait groups, the structured deme model predicts that to be favored by selection, the *A*-type must increase its own fitness in an absolute sense, not relative to others in the trait group. One could term this process "absolute individual fitness." It cannot be overemphasized how different it is from the "relative individual fitness" of the individual-selection model ($d > r$). In fact, most of the hypotheses developed in the following chapters rest upon equation (2.18), and we shall return to it repeatedly.

For this particular linear model, the difference between absolute and relative individual fitness is the class of traits for which $r > d > 0$, in which the *A*-type increases its own fitness but lowers it relative to others in the same trait group. Some confusion exists over what to call these traits. Theoreticians (e.g., Matessi and Jayakar 1976, Cohen and Eshel 1976, Eshel and Cohen 1976, Eshel 1977) tend to lump them with traits for which $d < 0$ and term them both altruistic, because they are selected against by individual-selection models. Nonmathematical treatments often speak of an altruist's "sacrifice" without specifying whether the sacrifice is absolute or is relative to others in the local population. I have been reluctant to term these traits "altruistic" (Wilson 1975a, 1977a) because in the intuitive sense of the word, traits that fall into this category do not appear altruistic and sometimes appear very selfish (see later chapters for examples). Nevertheless, in order not to deviate too widely from current terminology, the class of traits for which $r > d > 0$ should perhaps be termed "weak altruism" and the class of traits for which $d < 0$ be termed "strong altruism" (Wilson 1979). This is the usage adopted in this book.

Greater Than Bionomial Distribution. As trait-group variation exceeds the binomial, the bracketed term of equation (2.16) becomes negative, and traits that actually decrease the fitness of the individual performing them ($d < 0$, strong altruism) can be selected for, given a sufficiently positive effect on the trait group at large (r). Hence, the structured deme model predicts that the binomial distribution represents a threshold variance for the evolution of strong altruism. As trait-group variation goes from point 2 to point 4 of the continuum, increasingly stronger altruistic traits can be selected for. Natural selection becomes more sensitive to the effect of traits on the population at large.

Point 3, Kin Selection. Kin-selection theory explores the evolutionary effects of individuals on their relatives. In some cases, these effects are behaviorally directed (i.e., an individual recognizes relatives and behaves differently toward them than toward nonrelatives); in other cases, the effects are spatially directed (i.e., an individual cannot distinguish relatives from nonrelatives, but directs most of its behavior toward the former simply because of their spatial proximity). Spatially directed kin selection represents a special case of structured demes in which the kin group equals the trait group. Therefore, one should be able to calculate average subjective frequencies that are characteristic of kin groups and arrive at the same conclusions as classical kin-selection theory (Wilson 1977a).

Consider a sexual haploid population containing two alleles, A and B, in proportions p and q (different letters used to preserve terminology). Mating occurs at random, so the frequency of A-A, A-B, and B-B matings are p^2, $2pq$, and q^2 respectively, and

$p^2/(p^2 + pq)$ = the proportion of A-offspring resulting from A-A matings

$pq/(p^2 + pq)$ = the proportion of A-offspring resulting

from A-B matings
$q^2/(q^2 + pq)$ = the proportion of B-offspring resulting from B-B matings
$pq/(q^2 + pq)$ = the proportion of B-offspring resulting from A-B matings

Each female has a clutch of offspring (size N) that remains isolated from other clutches. Because interactions are restricted to within a clutch, each clutch constitutes a trait group, and each trait group is composed entirely of siblings.

The subjective frequencies for this situation may be calculated as follows: The clutches from A-A and B-B matings will have frequencies of $p = 1$, $q = 0$ and $p = 0$, $q = 1$, respectively. The clutches from A-B matings will have mean frequencies of $p = q = 0.5$ and a binomially distributed variance of $\sigma^2 = pq/N = 0.25/N$. Thus, the average A offspring from A-B matings will experience a subjective frequency of $p_A = 0.5 + 0.5/N$, while B-offspring from A-B matings will experience a subjective frequency of $p_A = 0.5 - 0.5/N$. The subjective frequencies for all offspring from all matings are then

$$p_A = \frac{p^2(1) + pq(.5 + .5/N)}{p^2 + pq} = p + q(.5 + .5/N) \quad \textbf{(2.19)}$$

$$p_B = \frac{q^2(0) + pq(.5 - .5/N)}{q^2 + pq} = p(.5 - .5/N) \quad \textbf{(2.20)}$$

These are the characteristic subjective frequencies when interactions occur exclusively among siblings.

Most kin-selection models deal with the relations between a donor (the A-type) and a single recipient. In this case, the effect of every A-type is divided among $(N - 1)$ recipients, and inequality (2.16) becomes

$$d + \frac{(Np_A - 1)r}{(N - 1)} > \frac{Np_B r}{(N - 1)}$$

$$d > \frac{r}{(N - 1)} \left[N(p_B - p_A) + 1\right] \quad \textbf{(2.21)}$$

Enter the values for the subjective frequencies as follows:

$$d > \frac{r}{(N - 1)} \left[N(p(.5 - .5/N) - p - q(.5 + .5/N)) + 1\right] \quad \textbf{(2.22)}$$

$$d > \frac{r}{(N - 1)} \left[N(-.5(p + q) - .5(p + q)/N) + 1\right]$$

$$d > \frac{r}{(N-1)} \left[-.5(N-1) \right]$$

$$d > -.5r \tag{2.23}$$

This corresponds to the well-known conclusion of kin-selection models: that to be favored by selection, the donor's cost must be less than one-half the recipient's gain, demonstrating the basic equivalence between kin selection and structured demes. (See also Hamilton 1975, Matessi and Jayakar 1976, and Eshel and Cohen 1976.) However, this does not mean that no difference exists between the two models (Maynard Smith 1976). The most fundamental aspect of structured demes is the existence of trait groups. Kin selection pertains only to a certain kind of variation among trait groups, given their existence. See Wilson (1979) for a fuller discussion of the distinction between kin selection, inclusive fitness, and structured demes.

Point 4, Total Altruism. At the opposite end of the continuum from point 1, types are distributed such that any one trait group consists either entirely of *A*-types or *B*-types; in other words, total segregation between types exists. This yields a value of $\sigma_p^2 = \overline{pq}$; entering into (2.16),

$$d > r \left[-N\left(\frac{pq}{p} + \frac{pq}{q}\right) + 1 \right] \tag{2.24}$$

$$d + (N-1)r > 0 \tag{2.25}$$

Notice that since the left-hand side of inequality (2.25) includes the effect of a single *A*-type on the entire trait group (the effect on itself plus the effect on the $N - 1$ recipients), satisfying the inequality by definition increases the net fitness of the group. Point 4, therefore, represents a form of pure group selection, in the sense that point 1 represents individual selection.

Point 4 is almost as absurd an extreme on the variation continuum as point 1, but sometimes the former can be found in nature. Pseudopopulations such as coral polyps meet the requirements, as do trait groups composed of single individuals. Less obvious is the fact that trait groups composed of two individuals can be totally altruistic when the individuals are monogamous parents of the same offspring. The sexual haploid model of point 3 yields trait groups composed of *AA*, *AB*, and *BB* parents. In every *AB* trait group, the fitness of each type is equal (because they literally have the same offspring), so any differences in fitness between the two types must arise from the *AA* and *BB* trait groups.

We can summarize the continuum of trait-group variation as follows: Individual selection models predict that to be favored by selection, an organism must have the highest fitness, relative to others in its trait group. In structured demes, natural selection becomes increasingly sensitive to the differential productivity of trait groups as the amount of genetic variation among trait groups increases. The criterion for the selection of the A-type passes from highest relative individual fitness within the trait group ($d > r$) to highest absolute individual fitness ($d > 0$), then through kin selection, and eventually to highest group fitness $[d + r(N - 1)] > 0$.

While the superficial criterion for selection shifts with the degree of trait-group variation in the structured deme model, it is crucial to remember that the same general question is always being asked: Which type has the highest relative fitness in the deme ($F_A > F_B$)? Relative fitness in the global population is one of the fundamental rules of evolution and can never be violated. It simply turns out that on point 2 of the continuum, for instance, the way to maximize relative fitness in the deme is to maximize absolute fitness in the trait group. What occurs in the trait group is not equivalent to what occurs in the deme. That is the difference between individual selection and the theory presented here.

A Graphical Approach

At this point it is desirable to introduce a way to visualize graphically the effect of structured demes on the evolution of traits (Charnov and Krebs 1975, Wilson 1977a). Figures 2.3(a)–(f) show various types of functions for f_A (solid line) and f_B (dashed line), with respect to the frequency of A. In Figure 2.3(a) the fitnesses are constants (frequency independent), and the B-type is selected for. This is the kind of obviously adaptive trait that occupied the thinking of early evolutionary biologists.

Figures 2.3 throughout (b) and (c) show the standard concept of frequency-dependent selection, leading either to a stable equilibrium [2.3(b)] or to selection for the most abundant type [2.3(c)] (Ayala and Campbell 1974).

Figures 2.3(d) and (e) show a different sort of frequency-dependent selection, seldom mentioned because it does not seem to lead to any interesting conclusions. In both cases the B-type is favored by selection, just as in Figure 2.3(a) (the lines need not be parallel). However, in Figure 2.3(e) selection against the A-type has reduced the absolute fitness of the population. Figure 2.3(e), in fact, corresponds to equation (2.2) where $r > 0$ and $d < r$. It represents the class of traits selected against by individual selection but selected for under certain circumstances in structured demes.

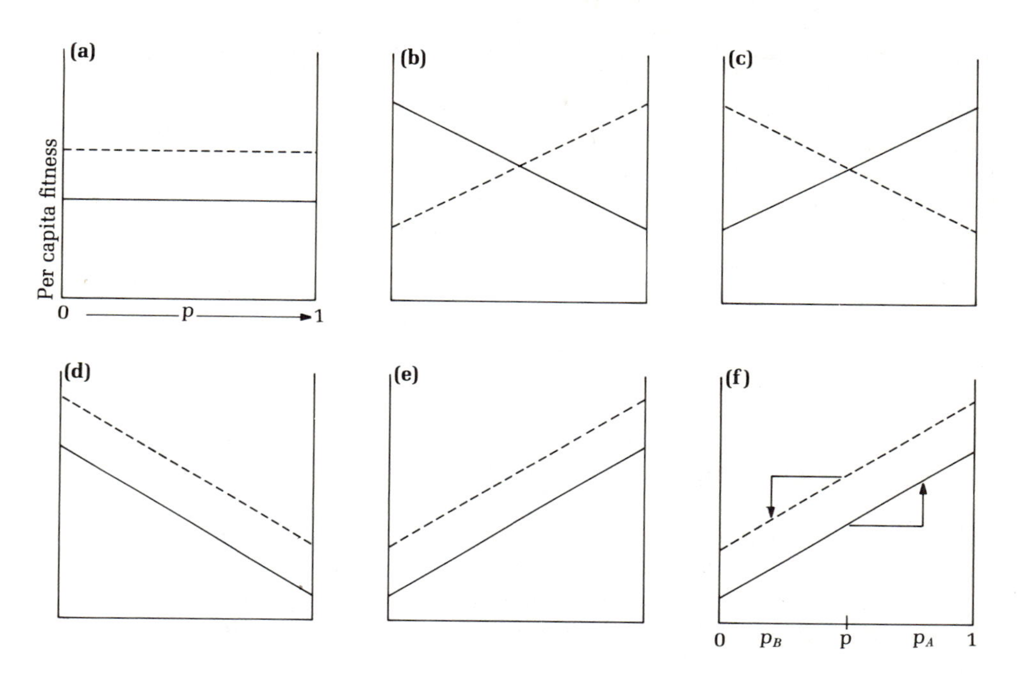

FIGURE 2.3 Per capita absolute fitness of *A* (solid line) and *B* (dashed line) types as a function of frequency. Total density (*N*) is held constant. Figure 2.3(a) = constant fitness values for each genotype. Figure 2.3(b) = frequency-dependent selection leading to stable, and Figure 2.3(*c*) = unstable equilibrium. Figures 2.3 (*d*) and (*e*) = frequency-dependent selection that does not lead to an equilibrium. Figure 2.3(*e*) = the concept of an altruistic trait, in which the *A*-type increases the fitness of the population (positive slopes) but nevertheless is always selected against (solid line always below dashed line). Figure 2.3 (*f*) = the effect of deme structure which under certain circumstances can reverse the outcome of selection, causing the altruist to be favored in evolution.

The effect of structured demes is shown in Figure 2.3(f). The subjective frequency and therefore per capita fitness is displaced to the right of the true frequency for the *A*-type and to the left for the *B*-type. This fact can reverse the outcome of selection—that is, it can cause the *A*-Type to have a higher per capita fitness than the *B*-type.

From Figure 2.3(f), it is easily seen that the selection of the *A*-type can be promoted in three ways: (1) The steeper the slope of the fitness functions, the smaller the displacement necessary to reverse the outcome of selection; (2) the smaller the vertical distance separating the fitness functions, the smaller the displacement necessary to reverse the outcome of selection; and (3) the larger the trait-group variation, the larger the displacement of the subjective from the true frequencies. Also, by applying the same process to the other graphs in Figure 2.3, it may be seen that deme structure has no effect on traits represented by constant fitness functions [2.3(a)] and little effect on frequency-dependent traits represented by Figures 2.3(b) and (c) (if $[f_A(p) + f_B(p)]/2$ is not constant, then the equilibrium point will shift slightly).

Although the use of subjective frequencies is quantitatively inappropriate, the graphical approach may also be used to explore the qualitative effects of nonlinear fitness functions. For instance, concave fitness functions can prevent the evolution of altruism at low frequencies, while convex functions may prevent its evolution at high frequencies, as rigorously demonstrated by Cohen and Eshel (1976).

The Distribution of Populations along the Gradient

So far we have demonstrated that the criterion for the selection of traits depends on the amount of trait-group variation. Now it is necessary to inspect where on the continuum of trait-group variation we are likely to find populations in nature. Figure 2.2 shows what I consider to be the probable distribution of populations with respect to trait-group variation, based on the following rationale: First, almost all structured demes should have a trait-group variation at or above the binomial distribution. (The most important exception would be demes in which the number of trait groups is small, in which case the variance in frequency is approximated by the hypergeometric distribution.) Hence, the distribution can safely be said to begin at point 2. Second, we know that a fairly substantial number of traits are closely approximated by kin-selection models. Hence, the distribution has not ended by point 3, although it should rapidly taper off above it. Finally, the peak of the distribution should occur slightly above point 2, corresponding to a slightly greater than binomial distribution (Wilson 1977a), caused by two processes. The first and most powerful process may be termed *diluted kin groups,* in which the offspring of several parents freely intermingle inside the trait groups. Consider an asexual population in which A and B represent two adult genotypes, distributed N to a trait group. If they are distributed binomially, then the variance in frequency is pq/N. Each adult now has e offspring which remain within the trait group. The mean and variance of offspring remain the same as for the parents, but had the offspring themselves been distributed binomially into the trait groups, their variance would have been pq/eN. The effect of reproduction inside the trait groups is to raise trait-group variation by a factor e over the random case. The concept of diluted kin groups holds for all organisms that do not disperse immediately upon release from the parent.

The second process for generating above binomial trait-group variation may be termed *differential mixing.* If the types interact differently with respect to the environment in almost any way, the differential interactions will translate into spatial differences and into an above binomial trait-group variation. As a hypothetical example, McAllister (1969) suggested that vertical migration of pelagic zooplankton represents a form of prudence to avoid the overgrazing of algae. (Many other plausible interpretations of vertical migration exist; see Enright 1977 for a review.) In addition, zooplankton that vary in their patterns of migration automatically expose themselves differentially to water currents, which translate into spatial horizontal segregation (Miller 1970). Thus, the same trait that is possibly group advantageous also alters spatial distributions.

In general, the binomial distribution is a statistical idealization which assumes that each trait group has an equal probability of receiving a given type. This assumption can be violated in so many ways that one shouldn't really expect random distributions to occur very often in nature (Wilson 1977a). Charlesworth (1978) has pointed out that in the context of a single locus model, the trait-group variation converges on the binomial as the frequency of either type approaches zero, in which case the differential mixing process could not explain the selection of strong altruism from a mutation frequency. However, the process may still be valid for the selection of quantitative traits, where rarity is not a problem (Wilson 1979). The whole concept of differential mixing deserves much more work.

Obviously the distribution in Figure 2.2 is very sketchy, and while an enormous literature exists on small-scale variation in allele frequency, it is of little use because of a powerful alternative explanation. Is the variation due to diluted kin groups and differential mixing, or is it the product of selection, due to differences in microhabitat (i.e., the multiple niche hypothesis, the alternative application of subdivided population models), a form of variation not applicable to the process investigated here? Both processes are likely to be pervasive in nature; it is difficult to separate the two (Wilson 1977a). Only one probable example of trait-group variation will be presented in this book. Istock and Weisburg (ms) studied allelic variation at two loci for the pitcher plant mosquito *Wyeomyia smithii* on a hierarchy of spatial levels: (1) Between pitchers. This corresponds to trait groups. (2) Between areas of the same bog. No significant differences were found, so the entire bog therefore constitutes the deme. (3) Between bogs. Pitcher plant mosquitoes are such weak fliers that negligible dispersal occurs between bogs. Hence, clusters of bogs constitute the "metapopulation" of traditional group-selection models (Levins 1970, E. O. Wilson 1975).

Table 2.1 shows the allelic frequencies for the phosphoglucomutase and phenyl-leucine amino peptidase loci in different pitchers within the same bog. Variation is considerable, probably due to the fact that females lay more than one egg in a single pitcher. This variation, therefore, is probably an example of diluted kin groups.

In order to concentrate on the most general aspects of structured-deme theory, it is necessary to be conservative and assume a trait-group variation that is either met or exceeded by most populations in nature. In my opinion, most populations may be safely assumed to lie at or above point 2 on the continuum of trait-group variation. In other words, we can conclude that weakly altruistic traits routinely evolve in nature, but that strongly altruistic traits require special mechanisms for high trait-group variation, and that the prevalence of these mechanisms is unknown. In order to retain the status of a general theory, most of the models in the following chapters assume the evolution of weak altruism only. As we shall see in Chapter 3, there is another good reason for following this conservative procedure.

Before leaving strongly altruistic traits, however, it is necessary to stress their potential importance in situations that are not often recognized. If differential mixing turns out to be a viable process, strong altruism may occur in populations for which it would never be expected on the basis of kin-selection theory. Furthermore, kin-selection theory itself has been applied predominantly to situations in which close relatives remain tightly associated with each other, to the neglect of diluted kin groups. If sibling insect larvae from a clutch of eggs are seen to mix freely with other clutches in their vicinity, it seems as if the conditions for kin selection have been violated, until one realizes that this mixing is trivial compared to that which occurs during a later dispersal stage. I have analyzed a possible example of altruism in diluted kin groups in detail elsewhere (Wilson 1977b).

Robustness

The preceding treatment of structured demes applies only to linear fitness models—in fact, mostly to a single type of linear model in which the fitness functions for the two types are separated by a constant interval. This model has the advantage of simplicity and will often be used in later chapters, but it lacks generality. As a

TABLE 2.1 Trait-group variation in the pitcher plant mosquito[a]

Plant	Pitcher	N	Phe-leu peptidase Allele A	Phe-leu peptidase Allele B	Phosphoglucomutase Allele A	Phosphoglucomutase Allele B
1	*A*	4	1.0000	0	.2500	.7500
	B	6	.9167	.0833	.2500	.7500
	C	2	.5000	.5000	.5000	.5000
	D	3	.6667	.3333	.6667	.3333
2	*A*	13	.6923	.3077	.3000	.7000
	B	2	1.0000	0	.7500	.2500
3	*A*	19	.7632	.2368	.3158	.6842
	B	3	.8333	.1667	.5000	.5000
4	*A*	19	.7895	.2105	.5263	.4737
	B	14	.7143	.2857	.5714	.4286
	C	11	.8182	.1818	.4545	.5455
5	*A*	10	.6500	.3500	.3000	.7000
	B	3	.8333	.1667	.3333	.6667
6	*A*	19	.7632	.2368	.3947	.6053
	B	20	.7750	.2250	.4500	.5500
	C	1	1.0000	0	.5000	.5000
	D	17	.8529	.1471	.5588	.4412
	E	10	.8500	.1500	.2778	.7222
7	*A*	19	.7632	.2368	.3421	.6579
	B	18	.8611	.1389	.4444	.5556
8	*A*	8	.8125	.1875	.5714	.4286
9	*A*	12	.7917	.2083	.3750	.6250
	B	12	.9167	.0833	.4583	.5417
	C	6	.7895	.2105	.1667	.8333
	D	18	.8889	.1111	.3889	.6111
10	*A*	5	.9000	.1000	.4000	.6000
11	*A*	18	.7778	.2222	.5833	.4167
	B	18	.8333	.1667	.5000	.5000
12	*A*	17	.7647	.2353	.6765	.3235
	B	17	.7353	.2647	.3529	.6471
13	*A*	18	.8889	.1111	.3056	.6944
	B	1	1.0000	0	.5000	.5000
14	*A*	17	.7941	.2059	.2647	.7353
	B	2	1.0000	0	.5000	.5000
	C	7	.7857	.2143	.3571	.6429
	D	11	.6818	.3182	.5000	.5000
15	*A*	19	.7368	.2632	.4167	.5833
	B	20	.7750	.2250	.2895	.7105
	C	8	.8125	.1875	.5000	.5000
16	*A*	20	.8250	.1750	.4750	.5250
	B	11	.6818	.3182	.3636	.6364
	C	20	.7250	.2750	.3947	.6053
17	*A*	20	.6500	.3500	—	—
	B	15	.8667	.1333	.5000	.5000
	C	2	.2500	.7500	0	1.0000

[a]Each plant contains one or more pitchers. *N* equals the population size of mosquitoes in each pitcher. Allele frequencies within each pitcher are given for two loci. Mean frequencies and average subjective frequencies for each loci are as follows:

$p = .7864$	$p_A = .7950$	$p_B = .7549$	$p = .4201$	$p_A = .4517$	$p_B = .4037$
$q = .2136$	$q_A = .2050$	$q_B = .2707$	$q = .5799$	$q_A = .5587$	$q_B = .6025$

further justification for accepting the concept of weak altruism as a conservative criterion for the natural selection of traits, it is desirable to consider the effect of structured demes for a more general class of fitness functions, and also to relax some of the assumptions of the structured deme model itself. The following treatment draws largely from the more mathematically rigorous treatments by Matessi and Jayakar (1976), Cohen and Eshel (1976), and Eshel (1977). Readers who desire only an intuitive understanding of the subject may skip to page 43.

Diploidy

Consider a single locus with two alleles, A and B (different letters used to preserve terminology). Let N_{AA}, N_{AB}, N_{BB} represent the density of the three genotypes in a trait group. With dominance, the fitness of the three genotypes within a single trait group is

$$f_{AA} = f_{AB} = d + (N_{AA} + N_{AB} - 1)\, r \tag{2.26}$$

$$f_{BB} = (N_{AA} + N_{AB})\, r \tag{2.27}$$

In structured demes with variable density between trait groups, we replace the densities with the average subjectives densities (2.12), which if $\text{cov}_{ij} = 0$ and $\sigma_i^2 = N_i$, leads to the following global fitnesses:

$$F_{AA} = d + (N_{AA} + 1 + N_{AB} - 1)\, r = d + (N_{AA} + N_{AB})\, r \tag{2.28}$$

$$F_{AB} = d + (N_{AA} + N_{AB} + 1 - 1)\, r = d + (N_{AA} + N_{AB})\, r \tag{2.29}$$

$$F_{BB} = (N_{AA} + N_{AB})\, r \tag{2.30}$$

It follows that the A-type will be selected whenever $d > 0$. A similar demonstration using recessive traits and constant trait-group densities is straightforward.

Continuous Trait Groups

The structured deme model assumes discrete trait groups—i.e., the organisms are gathered into discrete habitat units with homogeneous interactions within units and no interactions between units. This model accurately describes the structure of many populations, such as pitcher plant mosquitos, dung insects, birds in nests, and schools of fish. Another common type of trait group may be termed continuous, in which individuals are distributed relatively evenly over the deme, but each interacts only with its immediate neighbors. The set of organisms influenced by any individual's trait is still smaller than the deme (thus violating the homogeneity assumption of basic selection models), but each individual forms the center of its own trait group, and trait groups overlap with each other. Examples are territorial and sessile animals, and most plants.

Continuous trait groups are difficult to treat analytically, but easy to simulate on a computer. Accordingly, a 105 × 105 matrix was set up and "filled" with *A* and *B* types (Wilson 1977a). The types were entered at several overall frequencies (p = .05, .15, .4, .6, .85, .95) and in distributions both patchy and random with respect to each other. First the matrix was divided into discrete trait groups, and the subjective frequencies calculated. Trait groups were then constructed around every individual in the matrix to simulate the continuous case; again the subjective frequencies were tabulated. Trait groups could not be constructed around edge individuals, which were omitted from this study. In this analysis, three trait group sizes were used: 3 × 3, 5 × 5, and 7 × 7. In all cases, the discrete and continuous trait groups yielded identical results. Therefore, it appears that spatial variation in the genetic composition of populations is the fundamental parameter, regardless of whether or not the populations are gathered into discrete units.

General Fitness Functions

For a binomial distribution of genotypes into trait groups, the linear model employed so far gives rise to the pleasingly simple selection criterion of $d > 0$, which I have termed absolute individual fitness. Since many of my major conclusions rest on this concept, it is important to determine if it holds for a more general class of fitness functions. As we shall see, the $d > 0$ criterion is misleadingly simple in its apparent independence of such important parameters as density (N), frequency (p), and effect on recipient (r), yet the concept of weak altruism still retains a certain generality.

Following Matessi and Jayakar (1976), consider two types, *A* and *B*, in frequencies p and q over the entire deme. Within each trait group of size N, the fitness of each type depends on the abundance m of the *A*-type, i.e., $f_A(m)$, $f_B(m)$. With a binomial distribution of types among trait groups, the global fitness of the *A*-type is

$$F_A(p) = \sum_{1}^{N} \binom{N}{m} p^m q^{N-m} m f_A(m)/Np \tag{2.31}$$

But

$$\binom{N}{m} = \frac{N(N-1)\cdots(N-m+1)}{m!} = \frac{N}{m} \cdot$$

$$\frac{(N-1)(N-2)\cdots[(N-1)-(m-1)+1]}{(m-1)!} = \binom{N-1}{m-1}\frac{N}{m}$$

so (2.31) becomes

$$F_A(p) = Np \sum_{1}^{N} \binom{N-1}{m-1} p^{m-1} q^{N-m} f_A(m)/Np \tag{2.32}$$

$$= \sum_{0}^{N-1} \binom{N-1}{m} p^{m} q^{N-1-m} f_A(m+1) \tag{2.33}$$

$$= E\{f_A(m+1) \mid N-1, p\} \tag{2.34}$$

The global fitness of B-type is

$$F_B(p) = \sum_{0}^{N-1} \frac{N}{m} p^m q^{N-m} (N-m) f_B(m)/Nq \tag{2.35}$$

But $\binom{N}{m} = \dfrac{N}{N-m}$

$$\cdot \frac{(N-1)(N-2)\cdots(N-m+1)(N-1-m+1)}{m!}$$

$$= \binom{N-1}{m} \frac{N}{(N-m)}$$

and (2.35) equals

$$F_B(m) = Nq \sum_{0}^{N-1} \frac{N-1}{m} p^m q^{N-1-m} f_B(m)/Nq \tag{2.36}$$

$$= E\{f_B(m) \mid N-1, p\} \tag{2.37}$$

Similar conclusions were reached with a variable density between trait groups (Matessi and Jayakar 1976; see also Cohen and Eshel 1976).

Therefore, we can say in general that in structured demes with binomial trait-group variation, the global fitness of the B-type is equal to the expected value of its local fitness, while the global fitness of the A-type is equal to the expected value of its local fitness in trait groups to which a single A-type has been added. A sufficient condition for the selection of traits from mutation frequency to fixation is that $f_A(m+1) > f_B(m)$ for all m. Applying this general conclusion to the linear model [e.g., equation (2.2)] we immediately arrive at the $d > 0$ criterion. Other fitness functions are not as easy to partition into effects on self and effects on others, but it is obvious that since an A-type in a group with m + 1 A-types acts as a recipient to the same number of altruists as a

B-type in a group with m A-types, the A-type must have, in addition, a positive effect on itself to exceed the fitness of the B-type. To summarize, a binomial trait-group variation in structured demes will always favor a class of traits that are selected against in individual-selection models, and the concept of weak altruism holds for a broader class of fitness functions than the simple linear model used earlier.

Multiple Generations between Dispersal

Most organisms punctuate every generation with a dispersal stage. However, passive dispersers, such as some soil microorganisms, may go many generations before being mixed by physical forces, and other species may disperse only upon encountering the appropriate environmental cues, such as deteriorating resources. It therefore is desirable to consider the effects of multiple generations between dispersal periods. Such an analysis has been conducted by Cohen and Eshel (1976).

Of course, there is nothing magic about a single generation time. If fitness is defined as the rate of growth of the two types over the entire period of time spent within the trait group, then the criterion of selection determined by Cohen and Eshel (1976) for the binomial case is identical to that of Matessi and Jayakar (1976)—i.e., equations (2.34) and (2.37). Superficially, weak altruism is still selected for, regardless of how many generations are spent within the trait group. However, some traits that satisfy this requirement for a single generation may cease to do so after several generations. Indeed, since evolutionary forces within single trait groups act against weakly altruistic traits by definition, the B-type will always be selected, given a sufficiently long interval between dispersal (Cohen and Eshel 1976). Although the criterion for selection remains the same, the number of traits that fulfill the criterion diminishes.

The preceding considerations tend to support the routine selection of weakly altruistic traits when defined over the appropriate time interval. In addition, at least two situations may act against weak altruism. These are only mentioned here, a full exploration being beyond the scope of this chapter.

Delayed Benefit

The benefits of some traits do not occur until a considerable period after their manifestation, and these traits are a special problem for structured deme theory. Since the whole concept of structured demes relies on a spatial correlation between type frequency and trait-group productivity, that correlation is destroyed if dispersal occurs before the benefit manifests itself, and even weakly altruistic traits will be selected against. Notice that increased time spent within trait groups could actually improve the conditions for the evolution of altruism in these cases, contrary to the conclusions of Cohen and Eshel (1976).

One possible solution to the problem might be found by incorporating a stepping-stone dispersal process into the structured-deme model. It is well known that most populations do not occur within well-defined boundaries, with complete mixing upon dispersal. Instead, offspring disperse in a leptokurtic distribution around their parents, and their offspring do the same, so the average distance of a descendant from a given ancestor expands with each generation, like ripples spreading across a pond. As it stands, the structured-deme model requires that the set of organisms an individual interacts with (its trait group) is smaller than the set of organisms with which its offspring could potentially interact. However, if the set of organisms with which its grand-offspring interact is still larger, and so on, a series of nested trait groups could exist, each acting upon traits with increasingly longer time lags. Nonetheless, until such a process is accurately modeled, one should conservatively assume that natural selection is insensitive to traits whose benefits require many generations to manifest themselves.

Restricted Migration

Mathematically, the effect of restricting migration between trait groups is to weaken the conditions for the evolution of altruism (Charlesworth 1978). In my opinion, the small size of trait groups relative to the deme makes a restricted migration model inappropriate for most species. However, one important class of exceptions consists of vertebrates that live in tightly bounded social units (e.g., primate troops) with restricted movements between units. Ironically, these are the very species for which altruism is thought to be most prevalent.

The evolution of altruism in highly social groups nicely illustrates the necessity of determining the size of the local and the global populations for each and every trait. Many primate troops, for instance, are substructured into kin units. Trait groups for behavior in this context are equivalent to the kin group, with the entire troop approximating the deme (assuming a low dispersal rate between troops). However, many primate behaviors, such as choice of foraging areas and sleeping sites, affect the welfare of the troop at large. For these behaviors, the trait group equals the troop, and the effects of intertroop variation in genetic composition are best approximated by an island model. Finally, behaviors manifested during between-troup encounters bring more than one troop into a single trait group. The fact that the behaviors manifested within these three trait-groups sizes are almost certainly not independent of each other only complicates the problem. For the purposes of the following chapters, I make the assumption that the combination of kin selection, buildup of genetic variation between troops, and behavioral structuring meets or exceeds the conditions for the evolution of weak altruism.

Is It Group Selection?

Although Maynard Smith (1976) claimed that the structured deme model is no different from the model of kin selection, I more often encounter the argument that it is no different from the model of individual selection. Because relative fitness within the global population is still the criterion for selection, many people prefer to think of structured demes as a model of individual selection in which the homogeneity assumption is relaxed.

The distinction is largely semantic, but it is worth pointing out that exactly the same criticism applies equally well to traditional models of group selection. For instance, when the appropriate parameter values are chosen in Gilpin's (1975) model, selfish types eliminate altruistic types and then drive themselves to extinction, leaving empty habitats that are differentially colonized by dispersers from populations containing a high frequency of altruistic types—exactly as envisioned by Wynne-Edwards. The frequency of the *A*-type increases to fixation; i.e., it has the highest relative fitness throughout the global population.

To be consistent, one must either define group selection as that component of natural selection that operates on the differential productivity of local populations within a global system, or abandon the term altogether. There is already a trend toward renaming all forms of altruism that can evolve as "genetic selfishness," which presumably reserves the term "altruism" for anything that can't evolve (Alexander 1974). This conceptualization is a hollow victory for the individual selectionist.

Obviously, our theories on the evolution of social behavior are becoming rich enough that the relationships between formerly paradoxical elements are beginning to be seen, and a single model can produce the features of individual, kin, and group selection. Just because the relation between them is understood does not mean they need no longer be distinguished by combining them under a single term. To me, there are three strong reasons for retaining the term "group selection."

1. Observations on real organisms are almost always conducted within local populations. The concept of global fitness is fundamental but difficult to observe directly. In a practical sense we look at interactions within trait groups and need a terminology to describe the evolutionary forces that operate within trait groups. The term "individual selection" is appropriate.
2. Thinking of the between trait-group component of natural selection as group selection is profitable, and in many ways it is formally analogous to individual selection, as described in text books and as practiced in the laboratory. For instance, M. J. Wade is currently investigating the concept of group selection in the laboratory (Wade 1976, 1977). His experiments consist of creating a population of groups of *Tribolium* beetles that vary in their genetic composition, selecting groups on the basis of some criterion (e.g., high population size, low population size—although any measurable parameter would do), and using them to start a new generation of groups. He does with groups exactly what other geneticists do with individuals, and it would be difficult to avoid thinking of the process as group selection.
3. If the between trait-group component of natural selection has a powerful effect on the evolution of traits, then many organisms will be shown to behave in ways that maximize the productivity of the trait group, and, therefore, it is difficult to think of this as individual selection.

The various terms as I define them are summarized in Table 2.2. We may conclude by saying that certain features of the individual selection model, on which many evolutionary biologists base their understanding of intra- and interspecific interactions, are artifacts of a simplifying assumption of spatial homogeneity in the genetic composition of populations. When the homogeneity assumption is relaxed in a way that is biologically realistic for many species, natural selection routinely promotes the evolution of weakly altruistic traits, and also strongly altruistic traits when spe-

TABLE 2.2 Definition of terms

Individual selection	The component of natural selection that operates on the differential fitness of individuals within local (homogeneous with respect to genetic composition) populations
Group selection	The component of natural selection that operates on the differential productivity of local populations within a global population
Selfishness	All traits promoted by individual selection ($d > r$ in linear selection model)
Weak altruism	All nonselfish traits selected with a binomial trait-group variation ($0 < d < r$ in linear selection model)
Strong altruism	All nonselfish and nonweakly altruistic traits selected by a greater than binomial trait-group variation ($d < 0$, when r is sufficiently great, in linear selection models)

cial mechanisms exist to increase trait-group variation above the binomial distribution. Weakly altruistic traits are not spectacularly sacrificial, and at first do not appear to radically alter the conclusions of the individual selection model. Understanding their importance requires an analysis of the relation between group selection and altruism.

3 Group Selection without Strong Altruism

Traditional group-selection and kin-selection models seem to hinge upon proving the evolution of strong altruism. Ever since Wynne-Edwards (1962) stressed the individually disadvantageous nature of population regulation, altruism has served as the eye of the storm. Almost all discussions treat the disadvantage to the altruist relative to the nonaltruist as a measure of the strength of the process. Lack (1966) demonstrated that clutch size in birds tends to maximize the number of fledglings; he considered this conclusive evidence against group selection. Williams (1966) made altruism among nonrelatives a criterion for the demonstration of group selection. Even more recently, E. O. Wilson (1975) maintained that altruism is the "central theoretical problem of sociobiology."

If strong altruism is as necessary for group selection as currently believed, then structured demes do not constitute a very impressive theory (Matessi and Jayakar 1976). It predicts the routine evolution of only weak altruistic traits, and at its strongest, merely approaches kin selection.

Is strong altruism really necessary for a theory of group selection? I believe that it is not, and that the focus of the group-selection controversy has been off-center. The models discussed in Chapter 2 are sensitive to the group benefit of traits, anything that increases the relative productivity of local populations in a global system. A negative effect on the donor sometimes does not prevent the selection of such traits, but it always impedes it. To see this, consider our population of A and B types, and let each type possess a trait. The A-trait and B-trait are identical in terms of their effect on others in the trait group (r). But they differ in their effect on the individual performing the trait (d_A vs. d_B), in which case the fitnesses of the two types are

$$F_A = d_A + (N - 1)\,r \qquad F_B = d_B + (N - 1)\,r \tag{3.1}$$

and the A-type is selected whenever $d_A > d_B$. As simple as this conclusion is, it is one of the most sadly neglected facts of the group-selection controversy. It means that *even if altruism can be selected in nature, it will not be observed if there is a more selfish way to perform the same group benefit* (Gadgil 1975, Wilson 1977, West Eberhard 1975).

The most fundamental statement that can be made about group selection is that it tends to increase the productivity of trait groups. Altruism is a secondary consideration that becomes important only if a meaningful set of group-advantageous traits are for some reason necessarily altruistic. A rarity of altruism in nature may be explained in two ways: (1) the absence of the kind of population structure that can select for altruistic traits, or (2) the scarcity of group-advantageous traits that are necessarily altruistic. To put the matter even more forcefully, if all group-advantageous traits could be performed without any differential fitness between types within trait groups ($d = r$ in the linear model), then evolution would be sensitive only to the "between trait group" component of natural selection, and it would act to maximize trait-group productivity given any nonzero trait-group variation. Kin selection and other mechanisms that create high trait-group variation would increase the rate at which these group-advantageous traits evolve, but would not affect the fundamental outcome. If so, then by focusing on strong altruism, the entire group-selection controversy would have raged over a prediction that no consistent theory of group selection makes.

Although this is undoubtedly an overstatement, it is obvious that a careful evaluation of the importance of group selection in nature requires an understanding of the relation between altruism and group benefit. Let us begin where Wynne-Edwards did, with the observation that many activities which benefit the group, of necessity harm at least some individuals within that group. To me this observation is faultless, and applies not only to problems of population control in which reproductive activity or group membership is curtailed, but also to the question of roles in animal societies, in which individuals perform functions that differ in their effects on survival and reproductive success. Given the inherent sacrifice of many group-advantageous traits, the next question is, "Who sustains the sacrifice?" If the sacrifice is sustained by the individuals initiating the behavior (such as voluntary reproductive restraint), it is strongly altruistic. If the sacrifice is imposed upon other members of the group by the individuals initiating the behavior (such as one individual inhibiting the reproduction of another), it is not strongly altruistic. Whether this behavior is weakly altruistic or selfish will be discussed later in this chapter.

The existence of sacrifice, therefore, does not by itself require strong altruism. For strong altruism to play a role in the group-selection process, the sacrifice must *necessarily* be sustained by the individuals initiating the behavior. This necessary relationship may exist in some cases. For instance, an animal that spots a predator may have no choice but to expose itself, if it is to warn its group at all. However, many other group-advantageous traits, such as population regulation and role differentiation, do not seem to require self-sacrifice. Group selection may be a powerful evolutionary force in the selection of these latter traits, even with a low degree of trait-group variation.

This chapter attempts to show how group selection can operate without strong altruism. To do this, I have taken a single concept—optimal density—and treated it with a series of models, passing from the traditional idea of altruistic regulation to a more powerful form of regulation based on interference behavior. The chapter treats other provinces of group selection in a brief excursion into sociobiology.

First, it is necessary to discuss the concept of optimal density.

Optimal Density

If evolution on the population level exists, then trait-group productivity, or the proportional contribution of the trait group to the deme, is the analog to individual fitness, and it should be enhanced in nature. One of the prime determinants of productivity is population size. Usually an optimal density of organisms exists at which productivity is maximized (Figure 3.1). At lower densities, available resources remain unutilized, while at higher densities, a variety of overcompensating density-dependent forces occur.

The concept of optimal density is not just a theoretical construct—it has a solid empirical backing. Table 3.1 lists some studies that show an increase in population productivity from a decrease in population size, often caused by predators. These studies show only that an optimal density exists, not that the populations are regulated at the optimum.

TABLE 3.1 Productivity is maximized at an intermediate density, often caused by predators. See also reviews by Fujita and Utida (1953), Tanner (1966), Chew (1974), Ellison (1960) for grasslands, Barbosa and Peters (1970) for insects, and Backiel and LeCren (1967) for fish.

Organisms	*Reference*
Insect grazers increase productivity of hot-spring algae.	Brock (1967)[1]
Fish grazers increase productivity of coral reef algae.	Johannes et al. (1972)[1]
Planktivorous fish increase productivity of pelagic zooplankton.	Archibald (1975),[1] Grygierek (1962),[1] Grygierek et al. (1966),[1] Slobodkin and Richman (1956)[2]
Protozoan grazers stimulate bacterial productivity in aquatic systems.	Barsdate et al. (1974)[1]
Insect herbivores increase forest productivity.	Mattson and Addy (1975),[3] Rafes (1971)[3]
Leaf-cutter ants increase forest productivity.	Lugo et al. (1973)[1]
Vertebrate herbivores increase productivity of grassland.	Pearson (1965),[1] Vickery (1972),[1] McNaughton (1975)[1]
Limpet productivity is highest at intermediate density.	Branch (1975)[1]
Herbivorous fish increases productivity of algae.	Cooper (1973)[2]
Larval mosquito productivity is highest at intermediate density.	Barbosa et al. (1972)[2]
Deposit-feeding amphipod increases bacterial productivity at intermediate densities.	Hargrave (1970)[2]
Moth increases yield of turnips.	Taylor and Bardner (1968)[2]

[1]Field study.
[2]Lab study.
[3]Suggested but not conclusively demonstrated.

What are the determinants of optimal density? Several forces come to mind:

1. Prudence. If a consumer is to maximize the long-term yield of a self-renewing resource, it must harvest at an intermediate rate. It must leave a resource "capital" from which it can draw off "interest." This fact, long known to fisheries and wildlife biologists, was brought to the attention of pure ecologists by Slobodkin (1961, 1968) who termed it prudence. The idea became controversial when in 1962 Wynne-Edwards proposed that virtually all organisms are in potential danger of overexploiting their resources by eating up their capital, and offered his theory of group selection as an evolutionary mechanism whereby overexploitation is prevented.

While few ecologists accept Wynne-Edward's theory, many have agreed with him in assuming that individual selection tends relentlessly towards overexploitation. Even this is not necessarily true. Not only is there counter-evolution on the part of the resources (Rosenzweig 1973, Slobodkin 1974, Mertz and Wade 1976), but there exist at least two strong individual selective forces that oppose each other among consumers. The first force is *exploitation*, in which individuals increase their energy intake by increasing their ability at resource capture. This may lead to overexploitation of the resource. The second force is *interference*, in which individuals increase their energy intake by decreasing their competitor's ability at prey capture, through dominance, territoriality, and so on. This tends to counteract exploitation and may even lead to underutilization of resources.

To summarize, individual selection does not invariably lead to overexploitation. What can be said, however, is that individual

FIGURE 3.1 The concept of optimal density. The largest number of offspring (N_{t+1}) is produced by an intermediate number of parents (N_t).

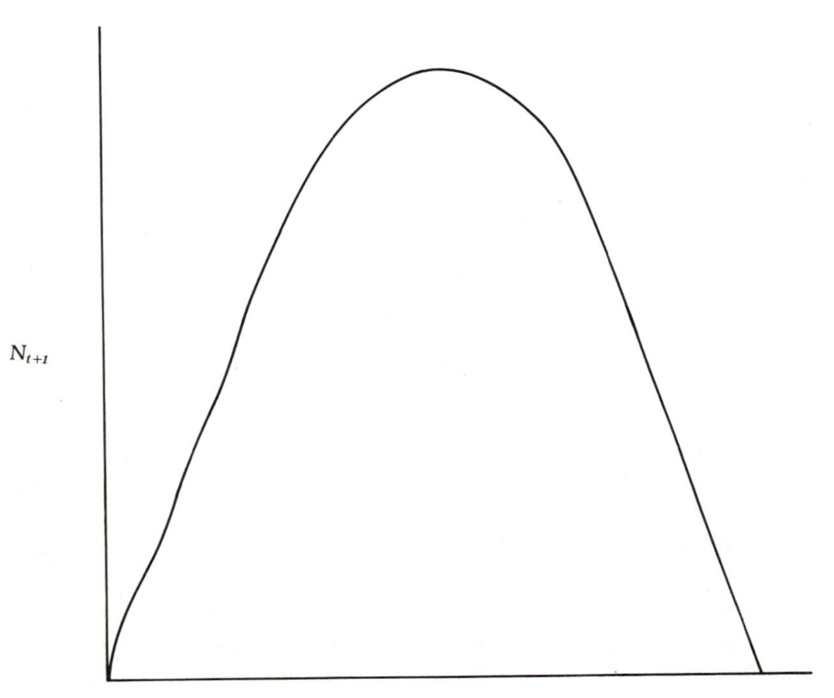

selection is insensitive to the optimal density. According to individual-selection models, equilibrium population size may be expected to drift aimlessly along the x-axis of Figure 3.1 through evolutionary time—rightward by exploitation, leftward by interference.

One other feature of Wynne-Edward's theory should be mentioned in passing. It has often been interpreted as a mechanism whereby populations are held at a submaximal productivity "to avoid overexploiting their resources." Actually, the question is more one of time spans—curtailing short-term productivity to enhance long-term productivity. All theories of group selection of which I am aware select for increased productivity of the local population.

2. Minimization of maintenance costs. Consider a stretch of lake shore, upon which 100 dead insects wash up from the surf every day. This is a resource that is impossible to overexploit. A species of carabid beetle feeds on these dead insects. Each beetle needs one insect a day for maintenance and can maximally feed on 5 insects a day. Everything above maintenance goes towards reproduction. If N is the density of beetles and h is a constant giving the conversion of food into offspring, then assuming the beetles can find all the insects, the population productivity equals

$$\begin{array}{lll} 4hN & \text{when} & N < 20 \\ h(100 - N) & \text{when} & 20 < N < 100 \\ 0 & \text{when} & 100 < N \end{array}$$

Productivity is maximized at $N = 20$ because above this density, energy is used in maintenance that could otherwise have been used for reproduction. In general, the optimal population size should consist of the minimum number of individuals that can utilize the available resources, thereby maximizing the energy devoted to reproduction. This concept should hold quite broadly for organisms, regardless of whether or not they are in a position to overexploit their resources. It also has several important corollaries. First, it suggests that the optimal population density is always below carrying capacity, where by definition most of the resources are consumed in maintenance. Second, this concept provides a general short-term advantage to population regulation, as opposed to resource exploitation, which may require several generations to manifest itself. Third, it suggests that if population regulation occurs, it will usually take the form of reducing population size, as opposed to each individual in the population reducing its consumption (or reproductive) rate. In other words, if food is the limiting factor, then regulation should occur by some individuals leaving the habitat entirely, with the rest reproducing as prodigiously as possible.

I know of two well-documented examples from nature. Redfield (1975) studied the population dynamics of the blue grouse in both increasing and stable populations. Reproducing females of both populations were found to be identical in mortality and fecundity statistics. Only the proportions of females allowed to breed differed between populations. Bryant and Hall (1975) were able to deduce the same fact in an elegant way for the housefly:

> *Musca domestica* L. is limited by cold weather during much of the year in temperate localities, but their enormous reproductive capacity allows their rapid increase to pest levels during favorable periods. At such high densities their contagious egg-laying behavior should lead to saturated oviposition sites, crowded larval conditions, and probable food limitation. Since crowding of muscoid larvae in the laboratory results in small adults, adult size in nature should reflect larval conditions. However, our extensive field observations on the housefly throughout the United States and those of Valiela (1969) on the face fly *M. autumnalis* (De Geer) reveal that diminutive flies of these species are rare, even when adult numbers are high. Hence, one is compelled to believe that larval populations are somehow regulated below their limiting densities. (See also Sokoloff 1955.)

Both of these studies suggest that population density was held at or below optimum, in terms of the minimization of maintenance costs. Of course, the big question is whether the regulation forms the adaptive value behind the trait, and we have no answer for that.

3. Nonresource influences. Resources are not the only determinants of optimal density. As only one example, a threshold density might exist above which disease epidemics occur. Optimal density would then surely exist below this threshold, regardless of the resource situation.

Now we are ready to study the evolution of density regulation in structured demes. In view of the mutiple factors affecting optimal density, most of the following models assume that population productivity follows a parabolic curve with respect to density (as in Figure 3.1) without assuming a specific biological interpretation (see Wilson 1977a for a more narrowly specific version). The equation used is

$$N_{t+1} = wN_t(Z - N_t) \tag{3.2}$$

where w and Z are constants. Optimal density for population productivity is at $N = Z/2$; trait-group extinction occurs at $N = Z$. Two additional features are of interest. First, note that absolute individual fitness equals

$$\frac{N_{t+1}}{N_t} = w(Z - N_t) \tag{3.3}$$

and declines continuously with population size. Hence, there is no special advantage to living in groups, and the optimal density for any given individual is at $N = 1$. Second, notice that the mathematical equilibrium occurs when

$$wN_t(Z - N_t) = N_t \tag{3.4}$$

$$N_t = Z - 1/w \tag{3.5}$$

Biologically, however, the population may be permanently maintained at lower densities (in particular, at the optimal density of $N_t = Z/2$) by a periodic cropping of excess individuals. In other words, an optimal density at $t = 0$ produces an above optimal density at $t = 1$, but excess individuals are then removed to recreate the optimal density. "Removal" is intended in the broad sense and includes quiescent stages in plants, insects, and other invertebrates; behavioral quiescence in vertebrates (nonbreeding by individuals capable of breeding); cannibalism; and relegation to habitats that do not support offspring (Brown 1969) or where exposure to predators is unusually high (Errington 1956). Most of the models that follow assume that the fitness of the removed individuals is zero. This is an exaggeration, but emphasizes the fundamental property we wish to explore, namely, the sacrifice of some for the benefit of the group. Group selection may be said to operate strongly when removal behavior evolves to regulate density at or near $N = Z/2$ regardless of whether the removal behavior is weakly or strongly altruistic. Individual selection is insensitive to the group optimum.

Exploitation

Voluntary Removal

The simplest and most stringent hypothesis is that a portion of the population voluntarily removes itself in response to overcrowding. Assume that the fitness of those that leave is zero; in other words, the sole advantage of the trait resides with those that are left behind.

Can voluntary removal be selected for in structured demes? Consider trait groups composed of two types, A and B, at a combined density of $N_t = N$ and in proportions p and q. Let trait-group productivity be governed by equation (3.2), and per capita fitness by (3.3).

Now assume that a portion of the A-types abandon the trait group—jump into the sea or walk into the mouths of predators—leaving a fraction k_1 remaining. This emigration creates a new population density,

$$N(q + pk_1) = N(1 - p + pk_1) = N(1 - p(1 - k_1)) \qquad \textbf{(3.6)}$$

and new per capita fitnesses, that are now different for the A and B types:

$$f_A = \frac{k_1 pN}{pN} w[Z - N(1 - p(1 - k_1))]$$

$$= k_1 w[Z - N(1 - p(1 - k_1))] \qquad \textbf{(3.7)}$$

$$f_B = \frac{bN}{bN} w[Z - N(1 - p(1 - k_1))]$$

$$= w[Z - N(1 - p(1 - k_1))] \qquad \textbf{(3.8)}$$

The per capita fitnesses of A and B are shown graphically in Figure 3.2(a). The (initial) presence of A does indeed increase the productivity of the trait group, but always to the differential advantage of B. Hence, emigration constitutes an altruistic trait, and is always selected against in individual-selection models. To see if altruistic removal can evolve in structured demes, we replace the average frequency with the average subjective frequencies:

$$F_A = k_1 w[Z - N(1 - p_A(1 - k_1)],$$

$$F_B = w[Z - N(1 - p_B(1 - k_1))] \qquad \textbf{(3.9)}$$

Let $y = F_A - F_B$

$$= k_1 Z - k_1^2 N p_A - k_1 N(1 - p_A) - Z + k_1 N p_B + N(1 - p_B) \qquad \textbf{(3.10)}$$

then

$$dy/dk_1 = Z - 2k_1 N p_A - N(1 - p_A) + N p_B \qquad \textbf{(3.11)}$$

and setting

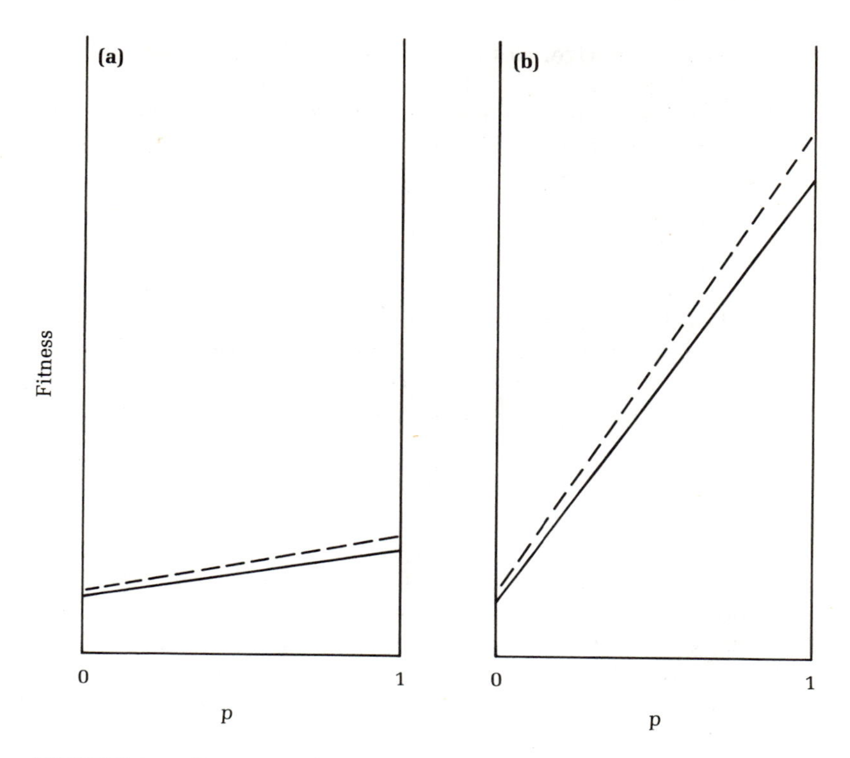

FIGURE 3.2 Per capita fitnesses for two models of population regulation. The solid and dashed lines are the fitnesses of the A- and B-types respectively. In (a), some of the A-types voluntarily leave the trait group, leaving a fraction k_1 remaining. In (b), the A-type adults feed upon a different resource, which allows them to produce only a fraction k_2 as many offspring as the B-type. The increased slope of the fitness functions in (b) enhances the conditions for selection of the A-type.

$$dy/dk_1 = 0, \quad k_1 = \frac{Z - N(1 - p_A - p_B)}{2Np_A} \tag{3.12}$$

Equation (3.12) is the optimal value of k_1, which maximizes the fitness difference between A and B. If $k_1 < 1$, then voluntary removal can be selected, at least under certain circumstances. (Alternatively, it can permanently exist as a trait that manifests itself only under certain conditions.)

$$\frac{Z - N(1 - p_A - p_B)}{2Np_A} < 1 \tag{3.13}$$

which is equivalent to

$$N(1 + p_A - p_B) > Z \tag{3.14}$$

To summarize, voluntarily removal can be selected in structured demes, but the conditions must be rather severe. For example, if the types are distributed binomially into trait groups, then $\sigma^2 = ab/N$ and equation (3.4) becomes

$$N(1 + p + \frac{pq}{pN} - p + \frac{pq}{qN}) > Z \quad (3.15)$$

$$N + 1 > Z \quad (3.16)$$

In this case the population must be on the very brink of extinction before voluntary removal takes place. Only when trait-group variation is on point 4 of the continuum (Figure 2.2, $\sigma^2 = ab$), does the population truly regulate its density at the optimum:

$$N(1 + p + \frac{pq}{p} - p + \frac{pq}{q}) > Z \quad (3.17)$$

$$N > Z/2 \quad (3.18)$$

This model, therefore, suggests that in structured demes voluntary removal may be observed in nature only under exceptional conditions, and is not a powerful regulatory mechanism. Models similar to this one give the impression that strong altruism, created by high trait-group variation, is necessary for group selection to act strongly in nature.

Competition between Age Classes

A small modification of the foregoing model is of interest, even though it does not increase the regulatory power of voluntary removal. All organisms exist in a range of sizes, from egg to adult. With the exception of those species that practice parental care, all stages in the life cycle must ecologically fend for themselves—yet they do so with varying abilities. Often the young form the weakest link in the population chain, inferior to older stages in both resource acquisition and susceptibility to predators.

This asymmetry can become especially critical when the different age classes compete for the same resources. Elsewhere I have argued that interstage competition might be a common event, in spite of large size differences between age classes (Wilson 1975b). That older animals often cannibalize smaller conspecifics as well as compete with them only adds to the dilemma. As one example of interstage competition, the larval ant lion's ability to capture prey largely depends on the circumference of its pit. I have estimated that the largest instar is from 13 to 50 times more efficient as a food getter than the smallest (biomass prey captured/biomass predator).

Furthermore, the largest completely overlaps the smallest in terms of prey size range (Wilson 1974). One can imagine a situation being generated in which one stage of the life cycle starves, while another is in the midst of plenty.

Consider a population of insects in which the adults compete with their larvae. Let the number of larvae surviving to adulthood be governed by the equation:

$$N_{G,t+1} = w_J N_{J,t}\left[Z - w_G N_{G,t} - w_J N_{J,t}\right] \qquad \textbf{(3.19)}$$

where N_G, N_J = the number of adults and larvae, and w_G, w_J = the resource capture rates of the adults and larvae.

As in our initial model, consider now an A-type and a B-type. A fraction of the A-adults disperse without producing offspring, creating a new density ($N_t = N$):

$$(N_G + N_J)(k_1 p + q) \qquad \textbf{(3.20)}$$

and new per capita fitnesses:

$$f_A = w_J k_1\left[Z - (N_G w_G + N_J w_J)(1 - p(1 - k_1))\right] \qquad \textbf{(3.21)}$$

$$f_B = w_J\left[Z - (N_G\, w_G + N_j w_j)(1 - p(1 - k_1))\right] \qquad \textbf{(3.22)}$$

Equations (3.21) and (3.22) are identical in form to equations (3.7) and (3.8), but with one important biological difference. Voluntary removal by the adult, when selected for, is caused by the conditions among the *larvae*. While voluntary removal does not function any better as a regulatory mechanism, it does suggest that a phenomenon might exist in nature that could never be explained by traditional models: the apparently maladaptive dispersal of individuals while in the midst of plenty.

Reducing the Sacrifice: Niche Differentiation between Stages

Although voluntary removal may be rejected as a powerful regulatory mechanism, this does not mean that removal itself cannot evolve through less altruistic pathways. In fact, even if voluntary removal were routinely selected in structured demes, one would not expect to observe it if less sacrificial ways existed to create the same effect. It therefore becomes important to look for ways to reduce the cost of removal. One possibility is niche differentiation. If the adult utilizes a different resource, it has removed its own competitive effect while still contributing to the larval population.

Consider two types of insects, A and B. The larvae of both types and the B-adults all utilize the same resource. The A-adults, however, occupy a different niche (another resource or perhaps a spatial/temporal separation). They thus do not compete with their larvae. However, the A-adults' niche is not as productive as the B-adults' niche. In particular, the A-adults produce only a fraction k_2 as many larvae.

Per capita fitnesses are represented thus:

$$f_A = w_J k_2 \left[Z - w_J N_J (k_2 p + q) - w_G N_G q\right] \tag{3.23}$$

$$f_B - w_J \left[Z - w_J N_J (k_2 p + q) - w_G N_G q\right] \tag{3.24}$$

These equations differ from equations (3.21) and (3.22) only in the absence of the A-adults' competitive effect on larval productivity.

We desire to know the minimum value of k_2 that still allows the selection for the A-type in structured demes, (i.e., the value of k_2 for which $F_A(p_A) - F_B(p_B)$ is slightly greater than zero). This value is difficult to arrive at analytically, but the numerical solution for one set of conditions is shown in Figure 3.3. The x-axis is $(w_J N_J (k_2 p + q) + w_G N_G q)$, or the effective population size, in terms of impact on the resource. Once again, the population goes extinct when the effective population size equals Z. In Figures 3.3 (a)–(c), N_G is variable,

$$N_J = 10 N_G,\ w_J = 1,\ w_G = 10, \text{ and } \sigma_p^2 = \frac{pq}{N_G}$$

In other words, the adults are distributed into trait groups according to the binomial distribution. Each adult has 10 offspring and is 10 times as efficient at acquiring the resources as a single larva.

In unstructured demes, $\sigma_p^2 = 0$ and the A-type is selected for only if $k_2 > 1$; that is, if the A-adult produces more offspring than the B-adult. However, in structured demes this situation is significantly altered. For instance, when $N_G = 5$, $N_J = 50$, and $p = 0.05$, an A-type that produces only 0.7 as many offspring as the B-type is favored by selection when the effective density is $0.75Z$—far below extinction levels. Adult densities as high as $N_G = 25$ still produce a significant departure from the individual-selection model. As the A-type increases in frequency, the conditions for its selection improve [Figures 3.3(b) and (c)]. Increasing the competitive efficiency of the adult relative to the larvae also enhances the effect of deme structure. (Figure 3.3(d), for which $N_G = 10$ and w_G is varied from 2.5 to 20).

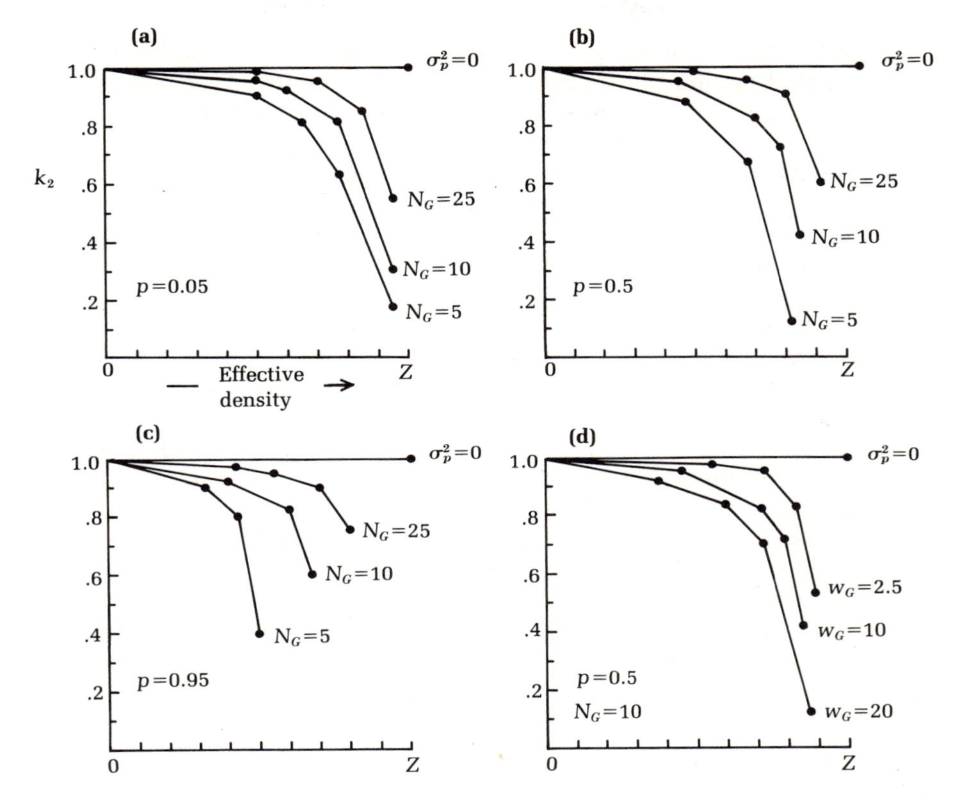

FIGURE 3.3 Evolution of complex life cycles when disadvantageous to the adult. k_2 is the fraction of offspring produced by the A-type, relative to the B-type within a trait group. The plotted curves are the minimal values of k_2 that allow selection of the A-type in structured demes. In (a)–(c), N_G is variable, $N_J = 10N_G$, $w_J = 1$, $w_G = 10$, $\sigma_p^2 = pq/N_G$. The frequency of the A-type is 0.05, 0.50, and 0.95 in (a) –(c), respectively. In (d), N_G is held constant at 10 and adult feeding rate (w_G) is varied from 2.5 to 20. See text for further explanations.

What has happened to produce this result, that makes niche differentiation so much more powerful than voluntary dispersal? Both involve altruistic traits, and in both cases the A-type produces only a fraction k of the offspring of the B-type. Per capita fitnesses of both types for the two models are shown in Figures 3.2(a) and (b) respectively. By removing its own competitive effect, the A-type in equation (3.23) increases the slopes of the fitness functions, thus enhancing its selection in structured demes.

Although some aspects of this model are artificial, its patterns are pronounced enough so that we can begin looking for possible examples in nature. Two will be discussed here.

King and Dawson (1972, 1973) raised flour beetles in a complex habitat containing two environmental gradients. From one side to the other of a square arena, the temperature varied from 22° to 34°C. Perpendicular to that gradient, the composition of the flour varied in four discrete steps, from 100% wheat to 100% corn. The animals showed age-specific preferences for points along these gradients. Adult *Tribolium confusum* occurred fairly evenly over the arena, but tended to lay eggs at a low temperature and at a high percentage of wheat flour. As the larvae matured, they migrated to increasingly warmer regions but remained within the wheat flour. Pupation occurred at high temperatures and in a high proportion of corn flour.

It is difficult to imagine on first principles why different ages should show such different preferences for temperature, and King and Dawson themselves suggest that it is an adaptation to avoid competition/predation between age classes: " . . . Indirect evidence indicates that females temporarily leave their home area to oviposit. This result may also be interpreted in terms of maximizing individual fitness in the midst of a 'cannibalistic orgy' because it suggests that the areas of egg laying and egg eating are spatially discrete." (King and Dawson 1972, p. 214). Laying eggs in safe places can indeed be explained in terms of individual selection, but cannibals avoiding areas of high egg density cannot. In this regard it is interesting that the eggs and the adult of *T. confusum* show a greater spatial separation than those of *T. castaneum* (King and Dawson 1973). Park et al. (1965) showed *confusum* to be the greater cannibal.

I have studied another possible example of niche shift that is individually disadvantageous to the adult in a tiger beetle (Coleoptera. Cicindelidae: Wilson 1978). All stages of tiger beetles are predatory and often feed heavily on ants (Balduf 1935, Larochelle 1974). The adult tiger beetle is highly mobile and runs swiftly about in search of its prey. The larva is an ambush predator. It digs a burrow and lies with its head flush with the surface of the ground, popping out like a jack-in-the-box to grab anything that comes within reach.

Cicindela repanda (Dejean), the most common North American tiger beetle, is often found on the narrow sandy margins of lakes, ponds, and rivers. This habitat is characterized by an absence of structural refugia for live prey, and allows extensive spatial overlap between tiger beetle stages. The flatness and openness of the terrain are especially favorable for searching predators, such as adult tiger beetles, and the difference in searching capacity between stages is in fact enormous. Ants observed on the beach in my study area encountered adult tiger beetles approximately once every eight minutes, but ants encountered tiger beetle larvae only once in four hours.

In contrast, five other congeneric species inhabit the fields and woodlands adjacent to the beach (*C. patruela, C. sexguttata, C. scuttelaris, C. punctulata, C. longilabris*). These habitats are characterized by a mosaic of patch types (e.g., sandy, mossy, rocky, hard vs. soft substrate), only some of which are suitable for larvae, and a large amount of structural refugia for prey in the form of leaf litter. These factors, combined with a low density relative to *repanda* (one could watch an ant all day without seeing an encounter with one of the inland species), reduce the possibility of interstage competition for these species.

When one observes the predatory behavior of adult tiger beetles in the field by walking quietly behind a foraging individual, the difference between *repanda* and the inland species is striking (Table 3.2). The inland species vigorously attack anything in the appropriate size range that moves. *Repanda* also shows a keen interest in movement, yet in two-thirds of all encounters with potential prey, it turned away without attacking, an interaction I term a "curtsy." *Repanda's* apparent cowardice has also been noted in the literature. Balduf (1935), citing an observation of Goldsmith (1916), states that while *repanda* has been observed to feed voraciously on small red ants, " . . . this species seems not as tiger-like in attacking larger and more active prey such as large black ants. The beetles usually make one dauntless charge, retreating suddenly if met by a counter attack, and give up in fear when repulsed." As a result, the adult *repanda* in my study area fed mostly on dead and dying insects washed up from the lake.

In order to determine whether small arthropods inhabiting the beach are suitable as prey for any tiger beetles, two inland species

TABLE 3.2 Observations of adult tiger beetles in their natural habitats. Parentheses give the proportion of encounters with live prey that led to curtsies, unsuccessful attacks, and captures. (From Wilson 1978.)

	Repanda	*Inland species*
Tiger beetles observed	60	22
Encounters with live prey	50	27
Curtsies	33(0.66)	2(0.07)
Unsuccessful attacks	15(0.28)	10(0.37)
Captures	3(0.06)	15(0.56)
Scavenging events	12	0

TABLE 3.3 Proportion of encounters with tiger-beetle prey that resulted in capture. Number of encounters are in parentheses. Prey size is given in the first column. (From Wilson 1978.)

Prey	Size(mm)	*Inland adults*	*Repanda adults*	*Repanda larvae*
Ants sp. [a,b]	2.7–6.7	0.65 (31)	0.04 (50)	0.39 (28)
Spider 1[a]	2.5	0.88 (16)	0.40 (10)	0.76 (17)
Carabid 1	6.14	0.06 (18)	0 (9)	0.06 (31)
Carabid 2	4.36	0 (12)	0 (20)	0 (3)
Carabid 3 [a,b]	4.22	0.93 (15)	0.07 (.4)	0.69 (26)
Carabid 4	1.54	0.17 (12)	0 (14)	0 (11)
Histerid 1	5.38	0 (7)	0 (13)	0 (6)
Collembolan1	1.2	0 (9)	0 (6)	0 (6)
Staphylinid 1	3.33	0.12 (17)	0.11 (9)	0 (7)
Dipteran sp.[b]	2.0–4.0	0 (6)	0 (14)	0.40 (12)

[a]Significant difference at or above 0.05 level for inland adults vs. *repanda* adults (X^2 contingency test).
[b]Significant difference at or above 0.05 level for *repanda* adults vs. *repanda* larvae (X^2 contingency test).

(*patruela* and *sexguttata*) were released into a large enclosure on the beach and observed. Feeding experiments were also conducted using small beach arthropods as prey. The results of these experiments showed that while many species have evolved effective defenses against all tiger beetles, others were easily attacked and eaten by the inland species, yet still refused by *repanda* adults (Table 3.3). My hypothesis is that the unique conditions of the beach habitat have thrown larval and adult *repanda* into intense competition with each other in a way that heavily favors the latter. There is little question that if adult *repanda* behaved as normal tiger beetles, the beach would soon be devoid of suitable prey. The adult stage has responded by shifting its niche in a way that is uniquely permitted by *repanda's* particular habitat. Of the two general categories of prey on the beach (living and dead/dying), both are available to the adult but only one to the larva, which is stationary and must rely on its prey to come to it. Merely by reducing the threshold of prey aggressiveness that inhibits attack, *repanda* adults become functional scavengers. It is difficult to quantify trait-group size for tiger beetles, but their movements make it likely that it is large. If my hypothesis is correct, then one would expect only a minor degree of sacrifice on the part of adults; i.e., they might refrain from living prey only when dead/dying food is abundant. This prediction has been confirmed. When held without food, *repanda* adults become predatory (Wilson 1978).

The most viable individual-selection hypothesis to explain the same data is that *repanda* adults have specialized in some way that prevents them from being effective predators. This possibility cannot be ruled out, but is not supported by the field study. In this regard, another species of tiger beetle currently being studied by D. L. Pearson (personal communication) is worth mentioning. The adults of this species, *C. willistoni,* congregate around the margins of shallow temporary saline ponds in Arizona, where dead insects washed upon the shores constitute three-quarters of their diet. However, upon encountering live insects, the adult readily attacks, showing no sign of restraint. The larvae do not occur on the pond margins, but rather in the ponds, where they extend their tunnels into towers that reach just above the surface of the water, and they catch flying insects that use the towers for landing pads. The habitats of *repanda* and *willistoni* are quite similar, but they differ in the extent to which they potentially compete with their larvae.

The difficulty of rigorously testing alternative hypotheses, even when the external patterns are as pronounced as they are with *repanda,* has been encountered again and again during the group-selection controversy. It is unlikely that any single field study can be so unambiguous and complete as to explore all options, especially since the sacrifice of group-advantageous traits will be minimized whenever possible and will be pronounced only when the ecological situation excludes other options. A convincing test, however, can still be accomplished through a large number of field studies on species that differ in interstage competition. Such a test would be conducted as follows:

1. Select a taxonomic group (e.g., a genus) in which the potential for interstage competition exists.
2. Divide this group into two subclasses: (a) those species for which the potential is realized by the ecological situation, and (b) those species in which the potential is not realized.
3. Both stages may be expected to shift their niche when individually advantageous. Presumably this may occur whether or not the stages compete. Structured deme theory predicts that, in addition, the adult niche may shift when advantageous to the larvae. This is only expected to occur in subclass (a). We therefore predict a higher frequency of niche shift in the adult when it competes with its own larvae. The frequency should increase with decreasing trait-group size. In a significant number of cases, introducing species from

subclass (b) into the habitat of subclass (a) should demonstrate a net advantage of the niche shift to the larvae and no net advantage to the adult. Coincidence and cryptic factors may explain the apparent lack of individual advantage in any single case, but not in a large number of cases that exist in the patterns predicted by structured deme theory.

Interference

Perhaps the most general pathway for decreasing the cost of population regulation is interference behavior. Whereas in exploitation, organisms deal exclusively with their resources, in interference they deal directly with competitors. The concept of interference thus embraces an enormous, but not ubiquitous, range of phenomena, including territoriality (in the broadest sense of any enforced spacing behavior), dominance, cannibalism, and chemical inhibition. I could not possibly cover all these subjects here, but fortunately excellent reviews exist (e.g., Colwell and Fuentes 1975). E. O. Wilson (1975) treats the more social aspects of interference, dominance, and territoriality. Fox (1975) discusses cannibalism in relation to population regulation. Chemical inhibition is reviewed by Whittaker and Feeny (1971).

Out of this diversity of behaviors we may perhaps retrieve two generalities, which are the subject of the following section:

1. All forms in interference derive their individual advantage from differential *suppression* of resource utilization. (In cannibalism, competitors simultaneously serve as a resource.)
2. Nonheritable differences between individuals feature prominently in nearly all types of interference. In particular, even though genetic variation undoubtedly exists in interference behaviors, it is the young and disabled of all genotypes that bear the brunt of the inhibition. The importance of nonheritable factors most certainly is true for dominance, territoriality, and cannibalism—and may also apply to chemical interactions. Especially among aquatic organisms, the degree to which a substance is taken into the body is likely to be a function of surface area, while its effect will be a function of volume. The substance should thus differentially affect small (young) animals, with a

high surface to volume ratio. No realistic model of interference can fail to take nonheritable factors into account.

Nonselective Interference

As a prelude to a more realistic model, consider two types, *A* and *B*, in a population whose productivity is governed by equation (3.1). The *A*-type practices a form of interference that may be termed "nonselective"; i.e., it inhibits itself as much as the *B*-type. This type of interference is not necessarily unrealistic. As examples, an animal might produce a toxin without being resistant to it, or a behavioral encounter might consume the time of both participants equally.

Effective population size and therefore per capita fitnesses depend on the presence of the *A*-type:

$$N_{t+1} = wN_t(1 - k_3p)(Z - N(1 - k_3p)) \quad \textbf{(3.25)}$$

$$f_A = f_B = w(1 - k_3p)(Z - N_t(1 - k_3p)) \quad \textbf{(3.26)}$$

where k_3 is the per capita inhibition of the *A*-type on the population. When $k_3 = 0$, the *A*-type has no inhibitory effect and equation (3.25) reverts to equation (3.2).

In individual-selection models, nonselective interference is a neutral trait, selected neither for nor against. This occurs because while nonselective interference definitely produces a change in fitness, it does not generate the *difference* in fitness required for individual selection to act.

To identify when the *A*-type is selected for in structured demes, we follow the same procedure as in the voluntary removal model (let $N_t = N$):

$$\begin{aligned} y = F_A - F_B &= (1 - k_3p_A)(Z - (1 - k_3p_A)N) \\ &\quad - (1 - k_3p_B)(Z - (1 - k_3p_B)N) \quad \textbf{(3.27)} \\ &= (Z - N + 2k_3p_AN - k_3p_AZ - k_3^2Np_A^2) \\ &\quad - (Z - N + 2k_3p_BN - k_3p_BZ - k_3^2Np^2_B) \end{aligned}$$

$$\frac{dy}{dk_3} = 2p_AN - p_AZ - 2k_3Np^2_A - 2p_BN + p_BZ + 2k_3Np^2_B = 0 \quad \textbf{(3.28)}$$

$$k_32N(p^2_A - p^2_B) = (2N - Z)(p_A - p_B)$$

$$k_3 = \frac{(2N - Z)(p_A - p_B)}{2N(p_A + p_B)(p_A - p_B)} \quad \textbf{(3.29)}$$

$$= \frac{(2N - Z)}{2N\,(p_A + p_B)} \tag{3.30}$$

The $(p_A - p_B)$ term in the denominator of equation (3.29) is "legal" only for nonzero values: by definition, a requirement met in structured demes and violated in individual-selection models.

Equation (3.30) is the optimal value of k_3, which maximizes the difference between F_A and F_B. Nonselective interference is favored by selection in structured demes when the optimal $k_3 > 0$.

$$(2N - Z)/2N(p_A + p_B) > 0 \tag{3.31}$$

$$N > Z/2 \tag{3.32}$$

Recall that $N = Z/2$ is the peak of the parabola, representing maximum population productivity (Figure 3.1). In sum, nonselective interference functions as a perfect regulatory mechanism in structured demes, curtailing trait-group size whenever it exceeds the optimum. It does this given *any* amount of trait-group variation greater than zero.

The reason why nonselective interference is so much more powerful than any of the exploitation models is shown graphically in Figure 3.4. Through nonselective interference the *A*-type has erased the difference between fitness functions, which instead are represented by the same ascending line. The trait is not altruistic—it is neutral in terms of individual selection and con-

FIGURE 3.4 Per capita absolute fitnesses for nonselective interference. Because the fitness of the *A*- and *B*-types are the same, any amount of trait-group variation causes the selection of the *A*-type.

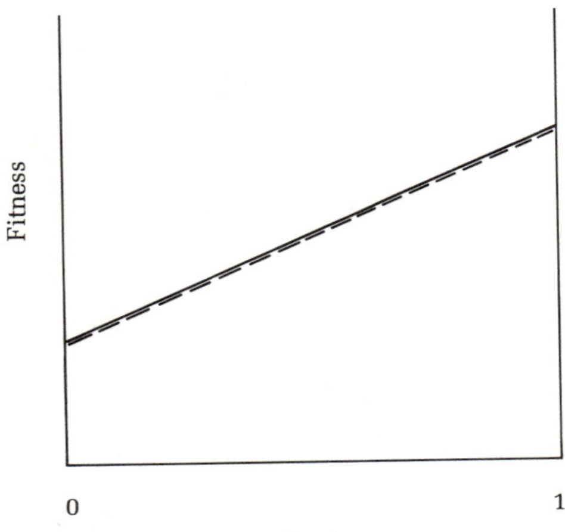

sequently needs only a tiny "nudge" by structured demes for its evolution. In essence, in exploitation the A-type is saying, "I'm going to curtail my density." In nonselective interference it says, "I'm going to curtail my density. . . *and so are you!*" The cost of removal is supported nondifferentially by all genotypes.

Selective Interference

There is, of course, no reason why interference must necessarily be nonselective. Our A-type could differentially inhibit either the B-type or itself. The former is required by individual selection, but the latter should not be excluded from our consideration.

As a beginning to exploring selective interference in a way that can be related to population regulation, only one small modification need be made in the previous model. Let k_4 = the per capita inhibition of A on itself, and lk_4 = the per capita inhibition of A on B. Density and per capita fitness then become

$$N_{t+1} = N_t\,(1 - p(pk_4 + qk_4 l)\,[Z - N_t\,(1 - p(pk_4 + qk_4 l))] \quad \textbf{(3.33)}$$

$$f_A = (1 - pk_4)[Z - N_t(1 - p(pk_4 + qk_4 l))] \quad \textbf{(3.34)}$$

$$f_B = (1 - pk_4 l)[Z - N_t(1 - p(pk_4 + qk_4 l))] \quad \textbf{(3.35)}$$

When $l > 1$, then A interferes more with the B-type; when $l < 1$, A interferes more with A-types. $l = 1$ corresponds to nonselective interference, and the equations revert to those in (3.26).

This model is fairly intractable analytically, but the major trends are explored graphically in Figure 3.5. First, it is obvious that the A-type is selected for in individual-selection models only when $l > 1$. In structured demes, however, the selection of A depends on the slopes of the fitness functions; these change with population size, as shown in Figure 3.5, for $N = 10$, 60, and 90. In Figure 3.5(a)–(c), $l = 0.75$, and in (d)–(f), $l = 1.25$.

At $N = 90$, interference is a group-advantageous trait because it moves the population toward the optimum. Therefore, the fitness functions in Figures 3.5(c) and (f) have a positive slope. In (f), the A-type is selected for and would be anyway in traditional models because it has the individual advantage. In (c), A constitutes an altruistic type and is selected for only given sufficient trait-group variation.

At $N = 10$, interference is a group-detrimental trait because it takes population size away from the optimum. Therefore, the fitness functions in Figures 3.5(a) and (d) have a negative slope. In (a), the B-type is selected for, as it always is in traditional models.

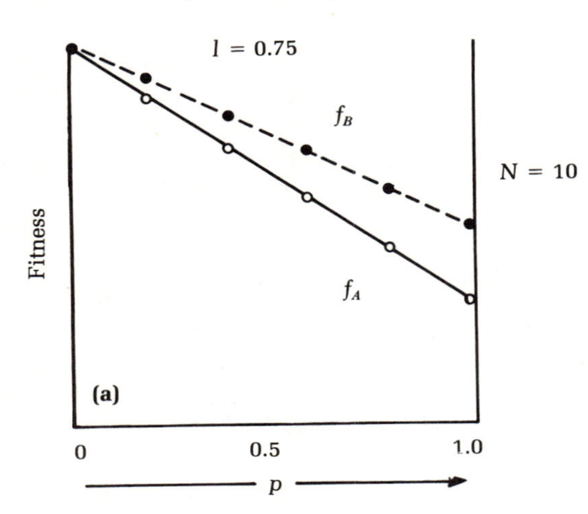
l = 0.75
f_B
f_A
Fitness
(a)
0
0.5
1.0
p

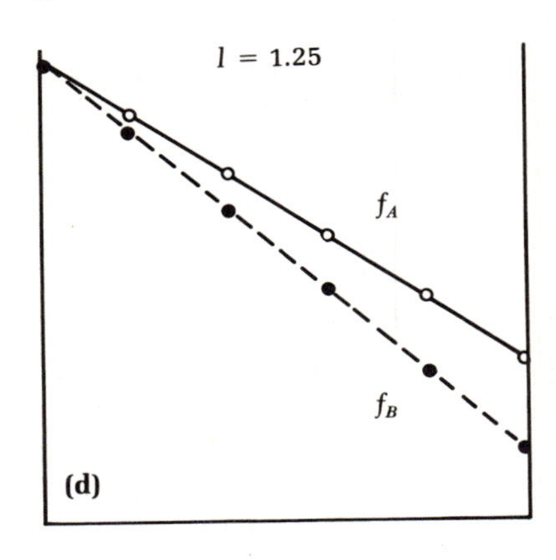
l = 1.25
f_A
f_B
(d)

N = 10

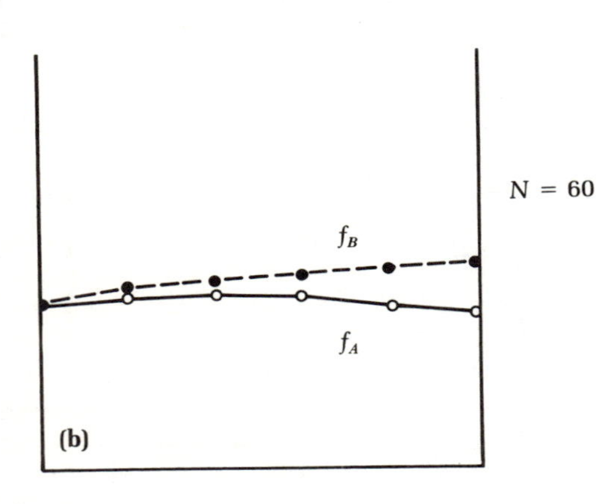
f_B
f_A
(b)

N = 60

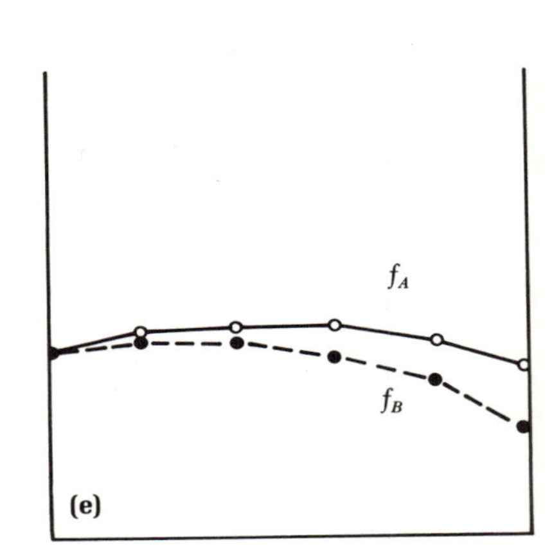
f_A
f_B
(e)

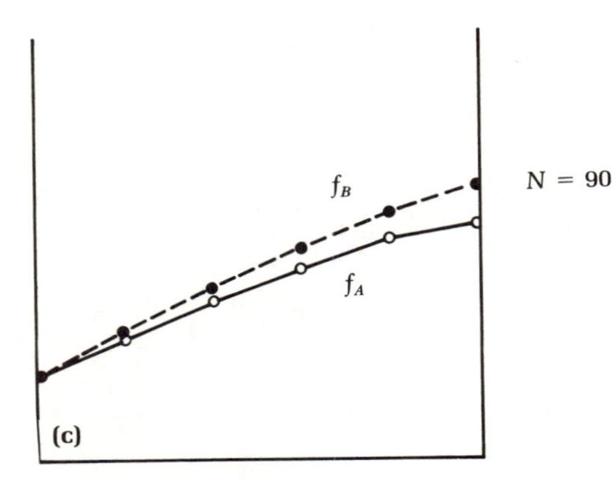
f_B
f_A
(c)

N = 90

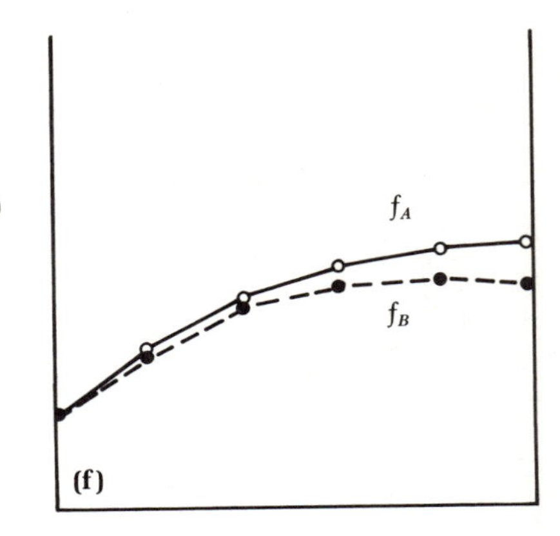
f_A
f_B
(f)

In (d), the B-type has become the altruist and is selected for over the A-type given sufficient trait-group variation.

In short, to function as a regulatory mechanism, selective interference must involve a measure of altruism, with either the A-type or B-type filling the role of altruist, depending on whether population size is above or below the optimum and which type is differentially interfered with. However, this kind of altruism differs from the altruism involved in voluntary removal in one fundamental respect. In exploitation the degree of altruism is rigidly related to the effect of the trait on population density. The term k_1 in equation (3.6) refers both to the decrease in population size and the sacrifice of the A-type. In the current model, k_4 refers only to the decrease in population size, while a separate term, l, governs the degree of sacrifice. There is no necessary relation between k_4 and l, which instead is determined by the specifics of the behavior.

Let us consider the regulatory properties of selective interference, using nonselective interference as a baseline. In Chapter 2 it was shown that the selection of an altruistic trait depends on three things: (1) the distance between fitness functions, (2) the slopes of the fitness functions, and (3) the amount of trait-group variation.

From our present standpoint, the most interesting of these factors are the slopes of the fitness functions, which are governed by density. Near optimum, the slopes are practically flat [Figures 3.5(b) and (e), $N = 60$]. The curves are bowed because the effective population size passes over the optimum as p goes from zero to one. At low densities the functions develop a negative slope [Figures 2.5(a) and (d), $N = 10$], and at high densities they develop a positive slope [Figures 2.5(c) and (f), $N = 90$]. The greater the departure of effective density from 50, the steeper the slope.

In nonselective interference there is no distance between fitness functions. Hence, any nonzero slope causes the selection of one type or another (or the appropriate response in an organism that can do both) and the population regulates its density perfectly. A small deviation of l from unity will create a small distance between fitness functions, and this distance will be counteracted by a gentle slope (i.e., a density near, but not at optimum). A large deviation of l from 1 creates a large distance between fitness functions, and this is counteracted only by a steep slope.

←**FIGURE 3.5** Per capita fitness for selective-interference model, as a function of frequency and density. Solid and dashed lines represent the fitness of A- and B-types respectively. In (a)–(c), the A-type differentially interferes with itself ($l = 0.75$). In (d)–(f), the A-type differentially interferes with the B-type ($l = 1.25$). Density (N) equals 10 in (a) and (d), 60 in (b) and (e), and 90 in (c) and (f). The slopes of the fitness functions are positive or negative, depending upon the density.

To summarize, selective interference systems do not lose their regulatory capacity; the point at which density is regulated merely shifts from the optimum. Small deviations of l from 1 cause small shifts, and large deviations cause large shifts. The direction of the shift is toward underpopulation when $l > 1$ and overpopulation when $l < 1$. Increasing trait-group variation always acts to bring regulation closer to the optimum. The central question about interference as a regulator in structured demes has become, "How selective is the interference?"

The Two Components of Interference

At this point it is necessary to inspect the concept of interference closely. I would like to suggest that it consists of not one, but at least two functionally independent traits:

1. The *vulnerability* of an individual to interference by others;
2. The patterns with which an individual *uses* interference against others (when able).

Perhaps some examples will help clarify the distinction between these two components:

1. An organism produces a toxin that inhibits population growth. The *use* of the toxin is governed by the existence of the gland, the amount of toxin synthesized, and the circumstances under which it is released. The *vulnerability* to the toxin is governed by quite a different set of factors, such as body size, an impenetrable integument, detoxification mechanisms, and so on.
2. Cannibalism almost invariably involves large animals preying on smaller ones, or on disabled individuals that are incapable of defending themselves (Fox 1975). Small and disabled animals are thus *vulnerable* to cannibalism. What, however, determines an animal's tendency to be a cannibal (the *use* of cannibalism)? The most common response to this question is "whenever an animal can get a meal out of being a cannibal." For the moment, however, it is necessary to ignore the adaptive genotypes and focus on the range of all possible genotypes, adaptive and maladaptive. In this sense, it is easy to imagine an animal capable of being a cannibal, that nevertheless is not. Of course, more subtle patterns of conditional

use are also possible, such as an individual that cannibalizes others' offspring but not its own, and so on.

3. As for cannibalism, it is easy to imagine an individual that is capable of interfering with others in aggressive encounters, yet does not. In other words, capacity for dominance by itself says nothing about the possible patterns with which dominance is used on others.

Maynard Smith and Price (1973), Maynard Smith (1974), Parker (1974), and Maynard Smith and Parker (1976) recognized the distinction between vulnerability and use of interference in their analysis of conflict between individual animals. These authors first define a "resource holding power" (Parker 1974—which may be interpreted here as vulnerability) and then examine a whole range of alternative behaviors (the use of aggression) for an evolutionarily stable strategy. Their approach presupposes that use and vulnerability may be considered independently, and that the level of vulnerability places no constraints on the patterns of use that can exist. The independence of use and vulnerability also acquires ample support from the empirical literature (King 1973, Gauthreaux 1978, Rowell 1974, Bernstein and Gordon 1974). However, some theoretical treatments (e.g., Gill 1974, Case and Gilpin 1974, Schoener 1976) do not distinguish between the two components, i.e., they model interference as a single trait, which assumes that dominant individuals necessarily exercise their power over subordinates.

The simplest two-trait model of interference would consider one gene with two alleles (*A* and *B*) that code for the use and nonuse of interference, and a second gene with two alleles (*C* and *D*) that code for vulnerability and invulnerability. This model generates four genotypes in haploids:

A/C, *A/D*, *B/C*, *B/D*

We specify that interference can be practiced only against individuals more vulnerable than the interferer. Thus, *A/D* individuals are invulnerable to interference, and themselves interfere with *A/C* and *B/C* types. *B/D* individuals are capable of resisting interference from *A/D* types, but themselves do not initiate interference against *A/C* and *B/C* types. *A/C* individuals are genetically disposed to interfere with others, but cannot in the absence of more vulnerable types.

The most obvious conclusion drawn from this model is that *A/C* and *B/C* types are at a constant disadvantage. Without some compensating benefit, vulnerability per se is selected against at all

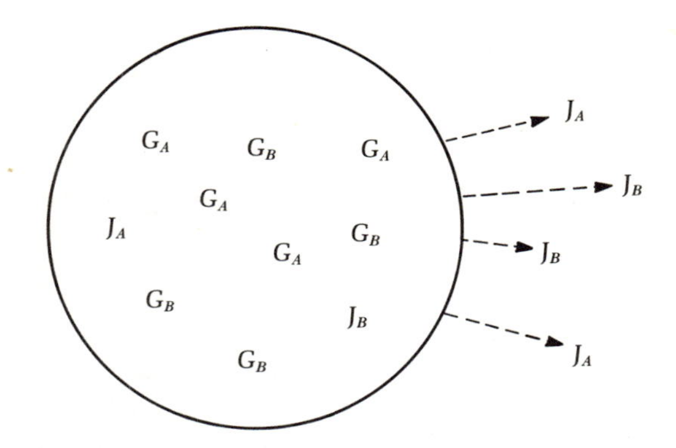

FIGURE 3.6 A trait group composed of two types (A, B) present as adults (G_A, G_B) and juveniles (J_A, J_B). A-adults expel individuals on the basis of their vulnerability, but this is genetically a nonselective event with respect to the proportion of A- and B-types remaining in the trait group.

times. Let us assume that this happens, leaving us with only two genotypes, A/D and B/D (relabel A and B), which differ in their tendency to use interference but not in their vulnerability to it. If no genotypic differences exist in vulnerability, how does the A-type practice interference? The answer is that even after the heritable component of vulnerability has been reduced to zero, there still exists a large nonheritable component in the form of age, sex, physical disability, and genetic load. These will combine to provide a phenotypic range of vulnerability in any population.

This simple model now exists in a form that can be related to density regulation in structured demes. Consider a trait group composed of two types, A and B, present as both adults (G_A, G_B) and juveniles (J_A, J_B). The division into adults and juveniles is intended as one kind of nonheritable variation in vulnerability. A-adults interfere with juveniles, causing the latter to leave the trait group. B-adults do not expel juveniles. The number of juveniles that leave are a function of the number of A-adults. The important implication is shown pictorially in Figure 3.6. Due to the presence of the A-adults, some of the juveniles are removed from the trait group. But because their expulsion is made on the basis of their vulnerability, they represent an unbiased sample of the trait group in terms of their inclusion of A- and B-types. Similarly, the lowered population size may have improved conditions in the trait group, but the improved conditions are enjoyed by A- and B-types alike.

In short, we have an event of great ecological importance that is phenotypically highly selective between individuals, yet it is neutral from the genetic standpoint. No change in gene frequency has occurred within the trait group. As such, the only evolutionary force producing genetic change is that which operates on the differential productivity between trait groups, caused by variation in the frequency of the *A*-type. Though phenotypically selective, this concept of interference is genetically nonselective and should be perfectly responsive to group selection in structured demes.

The most obvious way to increase the realism of this model is to consider the effect of an advantageous trait that increases vulnerability as a negative side effect. Assume a third gene with two alleles (*E* and *F*), one of which codes for such a trait, the other being neutral. The two traits again generate four genotypes:

A/E, *A/F*, *B/E*, *B/F*

Without even inquiring into the advantage of the *E* allele, it is obvious that it will have no effect on the model, since both its advantage and its negative side effect of increased vulnerability are conferred upon *A* and *B* types alike. Neither is linkage disequilibrium likely to have more than a transient effect on the outcome. Without linkage, selection for or against the *E* allele does not change the frequency of *A*, while total linkage is equivalent to a single gene, four-allele model. In the latter case there will still exist *A* and *B* types of equal fitness.

To summarize, a simple two-trait model predicts that many phenotypically selective forms of interference are genetically nonselective, and therefore are highly responsive to group selection in structured demes. One component of interference, use, selects against the other component, vulnerability, but selects neither for nor against itself. The only way to avoid this outcome is to demonstrate a necessary relationship between the two components, such that noninitiators of interference are automatically more vulnerable. Such a relationship may exist in specific instances, but I fail to see why it should hold in general. In fact, the only general relationship acts in the opposite direction, for *A*-types support an energetic cost of interference, while *B*-types do not. (Cost of interference is considered in more detail in Chapter 4.)

Not all kinds of interference fit realistically into this conception (see p. 82), but enough kinds do to warrant a few detailed examples. Once again, these examples do not constitute support for group selection—there are as many alternate hypotheses as one cares to think up. However, they do fit the assumptions with pleasing accuracy.

The Regulation of Parasite Infrapopulations

The concept of structured demes applies quite well to many species of parasites and disease organisms. For instance, the adult stage of the nematode parasite *Ostertagia* exists in the intestinal tract of livestock. The eggs pass out with the feces of the host, hatch, and develop through three larval stages in the external environment. The third stage larvae climb to the tips of grass blades where they are ingested by livestock, develop into fourth stage larvae, and finally, into adults (Olsen 1974). Ecological interactions can and do occur among individuals within a single host (the trait group). Yet the actual membership of each trait group consists of all third-stage larvae ingested within the grazing area of the host—offspring of parasites from many other trait groups (the deme). Parasitologists make the distinction between trait groups and demes, which they term "infrapopulations" and "suprapopulations" respectively (Esch, Gibbons, and Bourque 1975).

Although studies of parasites and density regulation have traditionally focused on the immune response of the host, it is becoming increasingly apparent that the parasites themselves possess considerable control over their own population dynamics. In *Ostertagia* and other nematodes, fourth-stage larvae are capable of entering a nonactive state of arrested development similar to the diapause of many insects. As in insects, arrested development in nematodes seems to serve a variety of functions. For instance, even though the internal environment of the host is relatively constant, the external environment fluctuates and, therefore, only some seasons are suitable for the development of the first three larval stages. Accordingly, during spring and summer the parasite infrapopulations consist largely of actively reproducing adult worms, while during the winter they consist largely of arrested larvae (Schad 1977).

Of greater interest from our standpoint is the possible use of arrested development for population regulation. In at least some species the number of arrested larvae appears to depend on the presence of adults in the infrapopulation. Several lines of evidence point to this conclusion. Gibson (1953) housed nematode-infested horses in conditions that prevented further infestation, and treated them with a compound that kills adult nematodes without affecting fourth-stage larvae. Each time the adults were eliminated from the trait group, a wave of larvae developed into adults. Michel (1963) infested several groups of calves daily with larval nematodes. In one group, adult worms were continually removed by treatment with a compound, while in a second group, adults were allowed to remain. Arrested larvae accumulated only in the group containing adults. Dunsmore (1963) conducted experiments similar to Gibson's with sheep, but sacrificed the animals soon after treatment to examine the immediate effect of adult removal on the development of the larvae. In this way, he convincingly demonstrated that removing the adults actually triggered larval development. (Schad 1977 reviews the subject of arrested development in nematodes.)

Although the presence of adults definitely influences the development of the larvae, the mechanism of the interaction is unknown. Several explanations are possible: (1) the larvae could "assess" the competitive environment and arrest development when advantageous to themselves, without any specially evolved behaviors on the part of the adult; (2) the larvae could altruistically arrest development when disadvantageous to themselves, to maximize the productivity of the infrapopulation; (3) the adults could inhibit the development of the larvae; or (4) the adults could modify the environment in some way (other than their normal feeding activities) to make it advantageous for larvae to arrest development. In this last case the larvae arrest development because it is advantageous to themselves, as in (1), but the "decision" is influenced by a specially evolved behavior on the part of the adult.

Situation (2) is a form of population regulation that involves strong altruism. Situations (3) and (4) accomplish exactly the same effect without being strongly altruistic. Neither are they selfish, because an adult nematode that arrests the development of larvae has not increased its fitness relative to other adults within the same host, and there is no reason to expect genetic differences between adults and larvae. Arrested development caused by adults is therefore a form of genetically nonselective interference. If adults are capable of arresting larval development, then the behavior should evolve to maximize the productivity of the infrapopulation (trait group), even with small amounts of variation in the genetic composition of parasite populations among hosts.

In general, I would like to suggest that parasites represent elegant organisms for the experimental investigation of the concepts developed in this chapter. They are one of the few groups for which trait-group productivity can be measured accurately. In addition, one can vary most of the parameters of the mathematical model (e.g., trait-group size, trait-group variation, number of generations between mixing, amount of mixing between trait groups) and find a natural system to match it. For instance, many parasites undergo parthenogenetic multiplication phases within their intermediate hosts, but exist as adults in their primary hosts, passing eggs into the environment without being able to multiply within that host. In a single species we have trait groups with high genetic similarity between members (intermediate hosts inhabited by genetically identical parasites) and trait groups with much lower genetic similarity between members (primary hosts inhabited by genetically diverse parasites). A comparison of their respective social behaviors will be fascinating.

Winter Flocking in the Harris Sparrow

The Harris sparrow *(Zonotrichia querula)* breeds on the arctic treeline, and in winter, migrates south to wide areas of the United States where it feeds in loose flocks. The regulation of flock size and the mechanisms behind it are the subject of an important series of papers by S. Rohwer and his co-workers (Rohwer 1975, 1977; Rohwer and Rohwer 1978; Rohwer and Winfield, ms).

First, it is necessary to ascertain the amount of trait-group variation in the system. The Harris sparrow is a nocturnal migrant, and any spatial proximity among kin on the breeding ground is likely to be lost during the night flights. Even after arriving at the winter feeding area, flocks are loosely organized, ranging over wide areas and continuously changing their membership. Thus, trait-group variation is likely to be quite low for the Harris sparrow, falling on or near point 2 of Figure 2.2.

In a separate study, Rohwer, Fretwell, and Tuckfield (1976) have actually provided an independent test of the level of trait-group variation. They noticed that when caught in mist nets and handled, some birds give off distress screams identical to those given when they are caught by predators. However, the frequency of this behavior varies with the species of bird: Some scream all the time and others not at all. Rohwer et al. reasoned that distress screams are intended to elicit an altruistic mobbing response from neighboring birds, and proceeded to relate the tendency to scream with the probable degree of kinship (trait-group variation). Sure enough, resident birds, who could be expected to possess close spatial proximity with siblings and offspring, showed a high frequency of screaming. Diurnal migrants could be expected to lose some (but not all) spatial proximity between kin during their daytime flights, and they screamed with intermediate frequency. Finally, nocturnal migrants could be expected to become almost completely scrambled during their night flights, and they showed a low frequency of screaming. In particular, only 1 out of 76 Harris sparrows tested emitted a distress scream.

By itself, this study is an intriguing demonstration of the dependence of altruism on trait-group variation. In terms of our present interests, it emphasizes that no form of population regulation in the Harris sparrow can rely on a large degree of altruism.

Rohwer came to study flock-size regulation while investigating a separate problem—plumage variability. Individual birds vary in their throat and breast coloration from almost pure white to totally black. Rohwer found that plumage is an accurate indicator of an individual's status in the group. Darker birds were dominant to lighter birds. Apparently the shifting nature of flock composition does not allow relationships to be worked out on the basis of encounters between individuals. Instead, a form of status signaling has evolved as an "instant relationship" whereby an individual can fit into a flock hierarchy immediately, without a lengthy period of challenges and rebuffs. The adaptive value of accurately signaling one's true fighting ability was shown in a series of experiments in which subordinate birds were dyed black to look dominant, and dominant birds bleached white to look subordinate (Rohwer 1977, Rohwer and Rohwer 1978).

Rohwer expected most of the aggressive interactions to occur among dominant birds in fighting for some sort of leadership position. However, the opposite, in fact, was found. Dominants chose to interfere mainly with the most subordinate individuals in the flock. A major class of exceptions to this rule consisted of injured birds (some of which were injured experimentally). They were treated as subordinates, regardless of signaled status.

More interesting was the form that the interference took. Dominant birds did not merely contest subordinates for individual items or positions within the flock. They actually chased them in a manner that drove them to the periphery of or completely away from the flock. Furthermore, the intensity of the interference fluctuated with the conditions, causing corresponding changes in flock size. For instance, during good weather little snow is on the ground, abundant seed supplies are available, aggressive interactions are few, and flock sizes are large. However, snow storms seem to be preceded by intense bursts of interference, and when the snow comes, causing a food crisis by burying most of the seed supplies, the flock sizes are small. Subordinates driven from the flocks are forced to exist as solitaries, in areas where food is too sparse or too evenly distributed to support the flocks at all (Rohwer, personal communication).

This is an example of population regulation based on interference, in which individuals are expelled on the basis of their vulnerability. It is obvious that vulnerability is strongly selected against. Why, then, do we see such variation in vulnerability? This question can have several answers. Vulnerability could be entirely nonheritable, due to age, sex, accidents, developmental abnormalities, or genetic load. Vulnerability, after all, is a close correlate of fitness in general, and a range of values is inevitable in any sexually reproducing population (Williams 1975). Alternatively, vulnerability could have hidden assets that manifest themselves in different situations, in which case it would be maintained as a polymorphism.

But the important point is that regardless of the answer to this question, it is the wrong question. The majority of interference studies focus on the fitness difference between the winner and the loser, when the true comparison must be between the winner and the equally invulnerable animal that has decided not to fight.

Rohwer noticed that not all of his dominant sparrows were equally aggressive. Some tended to chase subordinates more than others (personal communication). Consider an especially aggressive individual who is responsible for expelling most of the subordinates. If its aggression is timed to coincide properly with weather events, it may well increase the amount of food available for itself—but has it increased it relative to other, more passive dominants in the flock? It has not. If anything, the individual's relative fitness has been decreased by virtue of the energy expended in interference. (The cost of interference and other traits is treated in the next chapter.)

In summary, while vulnerability seems strongly selected against in the Harris sparrow, the use of interference appears to be close to neutral in terms of individual selection. I do not see how the use of this type of interference can be selected for on the basis of evolutionary forces operating within a single flock. As such, the strongest selection pressures molding the patterns of interference must be differential productivity between flocks (trait groups), in other words, group selection.

Population Regulation in a Ripple Bug (Hemiptera: Veliidae)

Wilson, Leighton, and Leighton (1979) have analyzed a remarkably parallel case in a completely different organism—a ripple bug (*Rhagovelia scabra* Bacon). The study is worth discussing in detail because it shows how interference can operate without a great deal of behavioral complexity, and also because it documents certain features about interference that are usually conjectural in the experimentally less tractable vertebrates.

Ripple bugs are similar to the stilt-legged water striders (Gerridae) in their ability to skate over the surface of the water. They are specifically adapted for life in the currents of streams, where they sometimes attain great numbers. Ripple bugs feed upon soft-bodied terrestrial insects floating on the water. However, unlike water striders, ripple bugs prey mostly on insects that are already dead. The population we observed did not attack struggling insects unless they were very minute (e.g., collembolans).

The nature of the habitat and the resource makes it possible to deduce the optimal location for food capture. If a ripple bug is in a

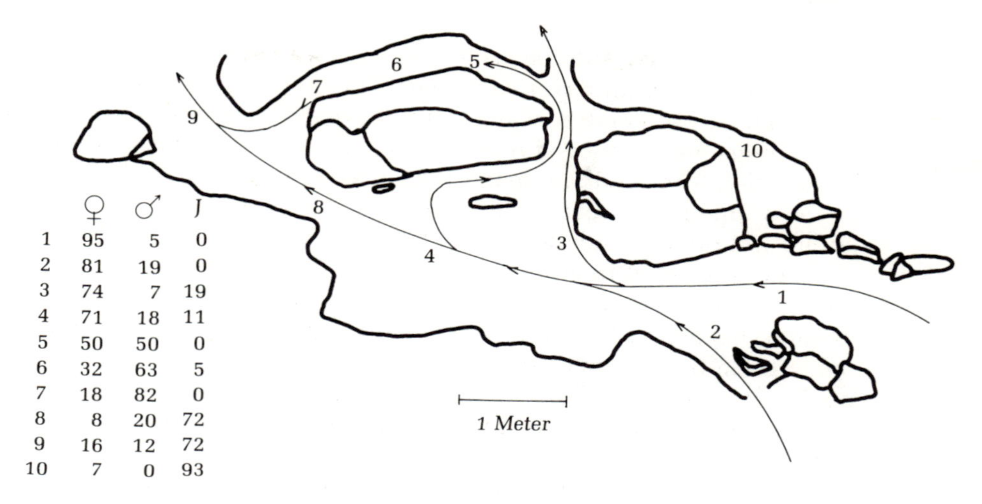

	♀	♂	J
1	95	5	0
2	81	19	0
3	74	7	19
4	71	18	11
5	50	50	0
6	32	63	5
7	18	82	0
8	8	20	72
9	16	12	72
10	7	0	93

FIGURE 3.7 A section of stream (drawn from a photograph) showing age/sex distribution of ripple bugs. The arrows represent the major flow of water. The numbers 1–10 represent sampling stations. The table gives the percent composition of adult females, adults males, and subadult instars (J) for each station. (From Wilson, Leighton, and Leighton, 1979.)

swift current (or better yet a quiet eddy bordering on a swift current), it will be able to survey a greater surface area of water flowing past than if it is in still water. In this way, areas of a stream can be ranked according to their resource availability.

When these areas were sampled for ripple bugs, a remarkable separation of age and sex classes was discovered. Figure 3.7 shows the spatial distribution of stages for one section of stream. Fast-flowing areas at the heads of pools invariably contained a high proportion of adult females. As the current became slower and slower, the composition shifted to a preponderance of males and finally to a preponderance of juvenile instars. The spatial separation was so pronounced that two adjacent sweeps of the net could be almost totally different in age/sex composition.

To see if these differences in position actually resulted in differential resource capture, we dropped dead fruit flies into the head of the pool (point 1, Figure 3.7) and recorded the time of their capture and the location and identity of their captors. The average fruit fly was intercepted in only 6.8 seconds, and 80% of the captures were made by adult females, 14% by males, and 6% by juveniles. Almost no fruit flies were carried to the tail of the pool. (However, this does not mean that ripple bugs at the tail of the pool did not feed—insects can fall into the stream at any point, and struggling insects can cease struggling at any point.)

Are the spatial differences between ages and sexes the result of interference or voluntary preference? This question, so difficult to approach in vertebrates, was easily answered in ripple bugs. All ripple bugs were removed from a pool, and 50 late instar juveniles introduced. They quickly made their way to the head of the pool, an area previously occupied only by adults. Adults were then added and premanipulation distributions re-emerged.

The process of interference was investigated in another way by creating a pool in which ages and sexes were color-coded. Fifty juvenile instars were marked with white enamel, 25 males marked red, and 25 females marked yellow; all were released into a vacant pool. Interactions, which took the form of chases and passive avoidances, could now be traced to the responsible sexes and instars.

Table 3.4 shows that interference operates along a vulnerability gradient. Juvenile instars yielded in encounters to females, males, and fellow juveniles; males (which are smaller than females) yielded only to females and males; and females yielded primarily to other females. Furthermore, juveniles were forced to yield a full four times more than females. As a result, females spent most of their time placidly gliding around the pool, while juveniles resembled molecules in brownian motion unless, of course, they could find an area uninhabited by adults. Finally, the enamel was poorly applied on some individuals, spreading over their legs and interfering with their movements in the water. These "injured" individuals were chased especially often.

In addition, dense clusters of ripple bugs existed in very still water along the edges of the stream (e.g., sampling station 10). These clusters were exceedingly uniform in their composition, being almost purely juvenile or male—but never female. There was

TABLE 3.4 The average number of times an individual ripple bug yields to another during a five-minute period (standard deviation in parentheses). Data are taken from three animals of each age/sex class. Females yield mostly to other females; males to females and males; and juveniles to males, females, and juveniles. Juveniles yield four times more than females. (From Wilson, Leighton, and Leighton 1979.)

		Yielded to			
		F	*M*	*J*	*Total*
Yielders	F	4.0 (2.7)	2.3 (0.6)	0 (0)	6.33
	M	10.3 (7.2)	3.3 (1.5)	0.7 (0.6)	14.33
	J	14.7 (5.7)	5.7 (2.9)	7.0 (3.6)	27.33

also a difference in behavior. Usually ripple bugs maintain a distance from their neighbors and are very active, even in still water. The animals in the clusters, however, were lethargic and packed leg to leg. Often a marked juvenile under observation would encounter one of these clusters in the process of being chased by adults. If the cluster consisted of fellow juveniles, the marked individual never left it.

Insofar as these clusters were located in the very worst areas in terms of resource capture, were far more densely concentrated than necessary, did not expend much energy, and consisted only of the more vulnerable age/sex classes, it seems likely that they represent individuals for whom the cost of interference is greater than the gain of finding food. They have "dropped out" of the population and must await a change in conditions to survive at all. We were fortunate to observe one such change in conditions: Toward the end of the study, heavy rains increased the water level of the stream by about four inches. Many areas that before were quiet, now had swift currents running through them—prime habitat for ripple bugs—and the clusters had largely disappeared.

The concept of optimal density in the ripple bug (and the Harris sparrow) is probably similar to the imaginary carabid beetle example on page 51. There is no danger of overexploiting the resource, but there is a density that can capture and consume most of the resource; above this density, energy is being used for maintenance that could have otherwise gone into reproduction. In the extreme case, densities can be attained for which there is not even enough food for maintenance, and the entire population starves together (a real possibility for the Harris sparrow after snow storms).

This situation, however, does not happen, and the reason is entirely due to interference. Because interference is applied differentially, habitats are not equally suitable for all members of the population. In particular, interference makes it individually advantageous for vulnerable animals to occupy suboptimal habitats or to stop feeding completely, even when resource conditions are optimal for those that remain.

Individual selectionists do not dispute this interpretation but maintain that it is a coincidence. The *real* selective force behind interference, they claim, is the advantage to the interferer. In the case of the ripple bug, this is equivalent to assuming that a female, for instance, who chases males and juveniles acquires something that more passive females in the trait group do not. There are two possibilities for such relative advantage: (1) a female could chase a vulnerable animal away from a favored position and then occupy that position herself; adjacent females would not benefit from this action; or (2) a female could have a chance at cannibalizing more vulnerable animals, thereby getting a meal as well as getting rid of a competitor.

We looked particularly closely for these phenomena but were unable to find any evidence for them. Many chases were prolonged and clearly designed to expel the individual from the general vicinity, not from a favored position. Cannibalism was not observed and is most unlikely, as *R. scabra* does not attempt to capture any prey of its own size, when it is still alive and struggling. It seems likely that the sole advantage of the trait lies in decreasing the population density of the trait group—an advantage shared by interferers and noninterferers alike.

Other Examples

I feel that the concept of nonselective interference is sufficiently realistic to predict that those populations meeting its assumptions regulate density to maximize the productivity of the trait group. Many flocking and schooling situations meet the requirements, as do parasites and disease populations. These assumptions might also be widely met in organisms that regulate their populations through chemical means (Whittaker and Feeny 1971).

Some readers will no doubt have noticed the strong similarity between the examples of interference presented above and the literature on population regulation through territoriality in birds (reviewed by Brown 1969, E. O. Wilson 1975). The latter studies also evoke the concept of optimal habitats, suboptimal habitats, and nonfunctioning floating populations whose occupancy depends on interference from the optimal habitat users. However, whether these cases fully meet the assumptions of nonselective interference is an open question. It is possible that territoriality falls within a large class of traits for which the individual selectionists' assumption is met, namely a differential advantage to the interferer. These forms of selective interference are discussed in the next section.

Selective Interference and Secondary Interference

The model of nonselective interference presented above has two essential ingredients. First, it is necessary that the most conspicuous feature of interference—the exclusion of a class of animals from its resources—is a genetically nonselective event as far as the use of interference is concerned. Second, the benefit of interference must be distributed among the remaining members of the trait group at large. These two ingredients make the entire act of interference neutral in terms of individual selection, and therefore perfectly responsive to group selection in structured demes.

For reasons already stated, I believe that the first ingredient is widely met, almost a necessary consequence of the nature of interference. The second ingredient is also commonly met, but it is not general. One can think of many examples in which the benefit of interference is not shared, but goes directly to the interferer. In a dispute over an individual item, the winner acquires the item. A cannibal not only lowers population size but feeds more than noncannibals. Indeed, whenever the profits of interference can be directed to the interferer, evolution may be expected to proceed in that direction. Thus, it is probably necessary for a resource to be indefensible to fulfill the assumptions of nonselective interference.

If the second ingredient is not met, then interference becomes genetically selective, and equilibrium population size deviates from the optimum. The extent of the deviation depends on how the interference benefits the trait group and how much of this benefit is differentially directed toward the interferer (Figure 3.5). Do any adaptations exist that can render this situation perfectly responsive to group selection again?

Consider a population in which selective interference is operating. Interferers get a large benefit relative to noninterferers—they are rapidly favored by selection and the equilibrium trait-group density declines to $N < Z/2$. Now consider an A-type who in some fashion is capable of interfering with the interference process, which may be termed *secondary interference*. The A-type's presence causes a new density, $N(1 + k_5p)$, and a new per capita fitness:

$$f_A = f_B = w(1 + k_5p)(Z - N(1 + k_5p)) \qquad \textbf{(3.36)}$$

Notice that equation (3.36) is identical to (3.26) with the exception that the A-type now causes an increase in density $(1 + k_5p)$ rather than a decrease $(1 - k_3p)$. Following the same procedure as for (3.26), it is easy to show that the A-type is selected for whenever $N < Z/2$.

To summarize, whenever selective interference leads to suboptimal density, it becomes itself the target for interference. Furthermore, by inhibiting another individual's actions, secondary interference is itself likely to be nonselective in its effect on the trait group. As such, the concept of secondary interference constitutes a pathway (theoretically at least) whereby populations capable of exhibiting selective interference may still be highly responsive to group selection in structured demes.

Candidates for secondary interference are not lacking. It obviously manifests itself in the arbitration of fights in certain primate troops (e.g., Eaton 1976). In a less obvious but probably more general way, a dominant male that allows females and infants closer than juvenile males creates a protective zone that limits the juvenile's capacity to interfere. In agonistic encounters, overtly aggressive behavior could elicit an even more violent counterresponse that would otherwise have been withheld. In this way an individual's behavior could be held within the limits tolerated by its neighbors, i.e., by secondary interference. A thorough analysis of secondary interference would be most rewarding but will not be attempted here. It should be mentioned in closing, however, that interference includes more than just agonistic behaviors. Female choice of mates, the granting and withholding of favors in relationships—anything that affects fitness and can be controlled by conspecifics falls into the same category.

An Excursion into Sociobiology

It should be clear by now that two pathways exist for the evolution of group-advantageous traits. The first could be termed the "altruistic" pathway because it assumes that the individual voluntarily sacrifices personal fitness for the welfare of the group. To evolve, these sacrificial traits require a high trait-group variation, and natural selection is exclusively responsive to group welfare only on point 4 of the continuum represented in Figure 2.2.

The alternative to the altruistic pathway may be called the "neutral" pathway. It requires that traits be neutral with respect to individual selection within trait groups, in which case they become perfectly responsive to the between trait-group component of selection in structured demes. Rather than relying on a large trait-group variation, the neutral pathway relies on a small vertical distance between fitness functions (Figure 2.3).

The concept of neutral traits sounds artificial and arbitrary when first encountered. An entire spectrum of social traits can be imagined, from highly advantageous to disadvantageous, and there is no reason to believe that a disproportionate number will be neutral. But the theory of group selection advanced here does not simply rely on neutral traits—it *generates* them through the use of interference against anything that is detrimental to the group, including interference itself.

In this section I would like to discuss briefly the implications of the neutral pathway for the study of group selection, and where we are to look for evidence of group selection in nature. Group selection pertains not only to the regulation of density or the overexploitation of resources. It also embraces all traits that influence the productivity of the trait group, including aspects of group organization newly gathered under the term "sociobiology" (E. O. Wilson 1975). Through the sociobiological literature runs a very strong current that distinguishes between the altruistic pathway and individual selection, while barely recognizing the existence of the neutral pathway (Gadgil 1975 provides the major exception). In only the latest expression of this trend, E. O. Wilson (1975) stresses that when considering the concept of roles in social groups, it is necessary to distinguish between those that are clearly designed to benefit the group and those that benefit the individual performing them. He then provides a long list of behaviors that are supposedly not material for group selection because they obviously benefit the individual manifesting the trait. For instance,

> *Fruit bats* (Pteropus giganteus) *form large daytime resting aggregations in certain trees in the Asiatic forests. Each male has its own resting position, with subordinate individuals occupying the lower limbs and hence suffering the most exposure to ground dwelling predators. The subordinate males usually see danger first and alert the remainder of the colony by their excited movements. They serve as very effective sentinels for the group as a whole (Neuweiler 1969) but their role is clearly indirect in nature (E. O. Wilson 1975, p. 319, underlining mine)*

Wilson then casts doubt on the entire concept of roles in vertebrate societies because most of them are precisely of this "selfish" kind. A similar sentiment exists in the literature on the helping phenomenon in birds, where a personal advantage to the helper is taken as evidence against kin selection (reviewed by Brown 1978).

By now it should be evident that these kinds of interactions among animals may preclude the altruistic pathway, but they say nothing about the possibility of group selection via nonsacrificial traits. Moreover, because the neutral pathway represents an evolutionarily stable strategy compared to the altruistic pathway, these patterns are actually expected. Let us elaborate on the example of sentry duty. What are the options? It is possible for an animal to march altruistically out to the edge of the group and scan the horizon for predators. Alternatively, interference can occur until the more vulnerable individuals in the group find it advantageous to sit at the edge, scanning the horizon for their own safety as well as that of others. Insofar as both accomplish exactly the same thing in terms of group benefit, one would never expect to see the former when the latter is possible.

But these are not the only options. Why is only enough aggression applied to force individuals to the edge of the group and not completely away, as in the Harris sparrow, where the danger comes from lack of food? Why is it that young males constitute the sentries, as opposed to equally or more vulnerable infants and females? In the fruit bat, females and infants are tolerated by the dominant males, a clear case of differences in the use of interference (Neuweiler 1969). If a young male attempted to force a more vulnerable female or infant to the edge of the group, thereby securing a safer place for itself, what would be the response of the older males? In the fruit bat, a young male cannot approach a female or infant without simultaneously approaching a dominant male, a clear case of secondary interference. There are obviously many ways to be selfish.

To determine the prevalence of group selection in nature, one must look to the balance of interference, the patterns of its use and counteruse. I believe that the altruism-selfishness controversy is peripheral to the question of group selection. Strong altruism may well exist in nature, and it may be important, but it is not the most fundamental part of any consistent group-selection theory.

The possibility of group-advantageous behavior without strong altruism has been recognized by several authors. As already mentioned, Gadgil (1975) was the first to reach the general conclusion. Lin and Michener (1972) and Michener and Brothers (1974) have cogently argued that the social insects are not as well explained by kin-selection theory as is sometimes claimed. As well as discussing cases where the assumptions of haplo-diploid models are not met (e.g., multiple insemination by males, one colony joining another when it becomes queenless), these authors provide many excellent examples of cooperation enforced by interference. (See also West Eberhard 1975). On a more speculative level, Trivers (1971), Alexander (1974), and Ghiselin (1974) have stressed that cooperation and supposed altruism contain ulterior motives or are enforced by interference. Often their arguments appear intended to reject the concept of group selection in general (especially Ghiselin 1974). In fact, they only reject the altruistic pathway.

One final interesting implication should be mentioned. If most group-advantageous traits can evolve through the neutral pathway, then high levels of trait-group variation, such as caused by interactions among kin, are not necessary for their selection. What then is the significance of kin selection for the evolution of social behavior? I do not have an answer to this question, but I would like to point out that the evidence for the importance of kinship in the evolution of social behavior is not nearly as strong as is often portrayed. The correlation between the two is undeniable, but the problem is one of causality. It is difficult indeed to imagine a close-knit society that does not generate interactions among kin as a consequence. On a finer scale, within such a society an individual may well structure its behavior so that beneficial effects are channeled toward kin whenever possible, yet this says little about the existence of the behavior in the absence of kin. In a similar way, many interactions among relatives can plausibly be explained as a consequence of social behavior rather than the cause of it. Undoubtedly, kinship does have important effects on social behavior, but its exploration must go beyond a simple correlation. Therefore, the study of those few social groups characterized by low relatedness between members takes on a special importance (e.g., parasites and winter bird flocks; see also the elegant study on bats by McCracken and Bradbury 1977).

To summarize, the distinction between group selection and individual selection is far more subtle than the altruism-selfishness dichotomy erected by Williams (1966—Williams' other cautions about the careless assignment of "functions" to traits applies just as forcefully, however). It will take some cleverness to separate the two experimentally. However, the difficulty brings with it a consolation: At best the altruistic pathway causes organisms to be only partially responsive to the benefit of the group. In the neutral pathway the potential exists for trait groups to truly maximize their productivity in nature.

4 The Evolution of Indirect Effects

Superorganisms

The concept of biological communities as "superorganisms" has arisen many times in the history of science. In this analogy a species is likened to an organ, whose behavior can be understood only in terms of its role in the maintenance of a larger whole.

With only a few notable exceptions (e.g., Dunbar 1960, 1972; Margalef 1968), modern evolutionary biology has been rather unsympathetic to the superorganism concept, largely because it seems irreconcilable with the Darwinian emphasis on individual self-interest (see Ghiselin 1974). In fact, given the limitations of the individual selection model, the two concepts really are irreconcilable. However, it is possible to show that communities in structured demes can evolve constraints so that the only way to maximize individual fitness is to perform in the interest of the community.

A Human Analogy

Like many other scientific terms, the word "community" already had a well-established everyday meaning when it was recruited for ecological use. Evidently the coiners of this word thought that human and biological communities have something in common. They are both collections of people/organisms, diverse in their activities, with a high degree of interdependence.

A brief inspection of human communities is worthwhile to see how far the analogy may be pressed. Few observers in our own society would doubt that a principle of "individual selection" operates strongly. Individuals act predominately to increase the welfare of themselves and their kin. The lawyer is, by and large, as selfishly motivated as the thief. Usually an individual is willing to help a common cause only if it does not involve too great a personal loss, and is most willing to help when he or she can simultaneously profit by it.

Yet at the same time it is impossible to avoid thinking of an individual as a cog in a machine, an organ dedicated to the maintenance of a much larger organism; and cynics notwithstanding, a human community is an organism of sorts, with its own physiology and laws of behavior, however poorly understood. In a sense, we really do exist for the function we perform in our community. Were there no demand for our function there would be no support for it, and we would promptly be forced to do something else: the "death" of one phenotype and the "birth" of another. An individual person may thus be equally well-construed as an independently operating entity or one completely dependent upon a larger whole. While this may at first appear paradoxical, it is in fact easily resolved.

Direct vs. Indirect Effects

It goes without saying that an individual person depends for his or her welfare on other people and their actions. No one has described the incredible interdependency of human communities better than Adam Smith (1776), in tracing the making of a pauper's overcoat:

> *Observe the accommodation of the most common artificer or day-labourer in a civilized and thriving country, and you will perceive that the number of people whose industry a part, though but a small part, has been employed in procuring him this accommodation, exceeds all computation. The woolen coat, for example, which covers the day-labourer, as coarse and rough as it may appear, is the product of the joint labour of a great multitude of workmen. The shepherd, the sorter of the wool, the wool-comber or carder, the dyer, the scribbler, the spinner, the weaver, the fuller, the dresser, with many others, must all join their different arts in order to complete even this homely production. How many merchants and carriers, besides, must have been employed*

> *in transporting the materials from some of those workmen to others who often live in a very distant part of the country! How much commerce and navigation in particular, how many ship-builders, sailors, sail-makers, rope-makers, must have been employed in order to bring together the different drugs made use of by the dyer, which often come from the remotest corners of the world! What a variety of labour too is necessary in order to produce the tools of the meanest of these workmen! To say nothing of such complicated machines as the ship of the sailor, the mill of the fuller, or even the loom of the weaver. Let us consider only what a variety of labour is requisite in order to form that very simple machine, the shears with which the shepherd clips the wool. The miner, the builder of the furnace for smelting the ore, the feller of the timber, the burner of the charcoal to be made use of in the smelting house, the brick-maker, the bricklayer, the workmen who attended the furnace, the millwright, the forger, the smith, must all of them join their different arts to produce them. . . . If we examine, I say, all of these things, and consider what a variety of labour is employed about each of them, we shall be sensible that without the assistance and cooperation of many thousands, the very meanest person in a civilized country could not be provided, even according to, what we very falsely imagine, the easy and simple manner in which he is commonly accommodated.* (Smith 1776, p. 11)

Add the more proximate forces of legal, economic, and social pressures, and this is the environment, characterized by enormous dependency, in which we operate. The dependency has an important and inevitable consequence. Consider an individual embedded in this framework, seeking to maximize personal gain. The person's actions can be divided into two components:

1. The direct effect of the act on the actor;
2. The response of the community to the act, which may be called the *indirect effect* of the act.

The indirect effect is a consequence of everyone in the community behaving exactly as the individual being considered, i.e., trying to maximize personal gain. For instance, the direct effect of a theft is the acquisition of a desirable item for the thief and its loss for the rest of the community. The indirect effect of the theft is the social and legal condemnation of the thief, if discovered by the community. Of course, indirect effects may also be positive. For example, many people dislike their jobs. The direct effects of their activities are negative to themselves, but positive to the rest of the community. The community responds with a positive indirect effect in the form of a salary. Notice that indirect effects by nature involve stimulating another component of the community, which then feeds back to oneself.

Individuals who desire to maximize personal gain must consider both the direct and indirect effects of their actions. In cases where the dependency of the individual on the community is weak, or where indirect effects can somehow be blocked (as for a thief who escapes capture), there is no necessary relation between the interests of the individual and the community, and they will often be at odds. However, if the dependency is strong—that is, if the community can exercise sufficient control over the individual (in both positive and negative ways)—then the indirect effects overwhelm and dominate the direct effects. The only way for the individual to further personal interests, therefore, is to act in the interests of the community. In this way the paradox between individual and community welfare is resolved.

I think it is evident that the concept of indirect effects is a major principle by which human societies operate. Its strength and pervasiveness would be hard to overestimate. In fact, if judged by their direct survival value, the vast majority of our activities would appear ridiculous. We live off the reaction of our society to our actions. A community without these indirect effects would be a very different community indeed.

Biological Communities

Biological communities are no different than human communities in the interdependency of their elements. One could describe a process at least as long as Adam Smith's on the species that influenced the chemical and structural composition of a clod of dirt. As we have seen, ecologists also possess the same dual concepts of individual self-interest on the one hand and the community superorganism on the other. However, traditional evolutionary theory treats them as an irreconcilable paradox. Natural selection cannot promote both—and, as we all know, it is individual selection that prevails. Ricklefs (1972) displays the current dogma nicely: "In terms of nutrient cycling in the ecosystem, microorganisms function to return the elements in organic compounds to inorganic forms, which can be used by plants; but bacteria and fungi decompose organic compounds for the same purpose that other organisms break down organic compounds—to release energy for useful metabolic work." Natural selection maximizes the fitness of individuals. The resulting traits may be beneficial or detrimental to the community at large, but the effects of the traits on the community are coincidental and have no bearing on their selection.

Let us see if this supposed paradox between individual and community advantage can be as easily resolved for biological as it is for human communities. No one denies that organisms in nature form a complex web of interrelationships. Mathematically, this web is usually represented by considering a number (*s*) of species, at densities $N_1, N_2, \ldots N_s$. The species are related to each other through a series of difference or differential equations, the change in density of each depending on the densities and the interaction coefficients of the others.

If the equations are linear, the interaction coefficients form the familiar community matrix, which in its most general form includes all types of species interactions. Linearity is useful as an illustrative tool but is not necessary for the general argument presented here:

$$\begin{bmatrix} a_{11} & a_{12} & \ldots & a_{1s} \\ a_{21} & a_{22} & \ldots & \\ \cdot & & & \\ \cdot & & & \\ \cdot & & & \\ a_{s1} & a_{s2} & \ldots & a_{ss} \end{bmatrix} \tag{4.1}$$

Here a_{ij} refers to the direct per capita effect of species *j* on species *i*. Indirect effects (for example, the effect of *j* on *i* through its effect on *h*, or the effect of *h* on itself through species *i*) are obtained by iterating the equations for more than one time interval. Any row of the interaction matrix gives the effect of the community on the species, whereas the corresponding column gives the effect of the species on the community.

We desire to explore the evolution of elements in the community matrix. One simple way to do this in a haploid model is to consider intraspecific variation as consisting of two types, represented by the first two rows and first two columns of the community matrix. The types are largely similar, but differ in one or more of their matrix elements. However, as a prelude to more interesting models, let the types start as identical:

$$\begin{bmatrix} g & g & h & i & j & \ldots \\ g & g & h & i & j & \ldots \\ k & k & a_{33} & a_{34} & \cdot & \ldots \\ l & l & a_{43} & \cdot & \cdot & \ldots \\ m & m & & & & \\ \cdot & \cdot & & & & \\ \cdot & \cdot & & & & \end{bmatrix} \tag{4.2}$$

Here type A and type B are represented as the first and second rows of the community matrix. The other species in the community are assumed to be monomorphic and occupy the rest of the matrix. The A- and B-types are identical with respect to their interactions with the community, represented by identical rows and columns in the matrix. Natural selection is obviously incapable of acting upon these types because there is no variation for it to act upon. If this model is run, the two types will simply retain their initial density ratio (N_A/N_B).

That natural selection needs variation in order to act is the first rule of evolution. Variation is ubiquitous in nature, so it is reasonable to expect our types to differ slightly in any or all of their coefficients, either in their rows or in their columns. First, consider a slight variation in the rows:

$$\begin{bmatrix} g & g & h & i & j & \dots \\ g & g & h' & i & j & \dots \\ k & k & a_{33} & a_{34} & . & \dots \\ l & l & a_{43} & . & & \\ m & m & & & & \\ . & . & & & & \\ . & . & & & & \end{bmatrix} \qquad (4.3)$$

Here, the two types are identical except for the effect of species 3 (h vs. h'). In this case, type A will be favored by selection whenever $h > h'$. For instance, if h and h' are per capita feeding rates on a prey population, then the one with the higher feeding rate is selected. When the model is run, one drives the other to extinction regardless of their starting densities. Variation among rows of the community matrix is in fact the evolutionary process as it is traditionally taught, and it corresponds to the concept of direct effects as used in the human analogy—what the organism extracts from its community.

Now consider a slight variation in the columns of the two types, that is, in their effect on the community:

$$\begin{bmatrix} g & g & h & i & j & \dots \\ g & g & h & i & j & \dots \\ k & k' & a_{33} & a_{34} & & \\ l & l & a_{43} & & & \\ m & m & & & & \\ . & . & & & & \\ . & . & & & & \end{bmatrix} \qquad (4.4)$$

Here the types are identical except for their effect on species 3 (k vs. k'). In spite of the existence of variation, natural selection is incapable of discriminating between the two types. If this model is run, their density ratio maintains its starting value, exactly as in the case where the types were truly identical. Mathematically, the reason for this conclusion is obvious. Identical rows in the matrix signify that the equations governing the fitness of the two types are identical, so a change in relative abundance is impossible. In this respect the columns do not matter.

This is not to say that a type's column in the community matrix has no effect on its fitness. On the contrary, it can have an overwhelming influence. In fact, every effect of a type on its community will theoretically come looping back to affect that type, either positively or negatively, via all possible pathways—a pathway being a chain of species connected by nonzero interaction coefficients. Column elements are, in fact, the concept of indirect effects as used in the human analogy—the stimulation or inhibition of another member of the community, which feeds back to affect one's own fitness.

The reason that natural selection is incapable of acting on variation in the column elements is not that they have no effect on fitness, but that they do not have a differential effect. A type cannot influence itself differentially by its influence on the community at large. All indirect effects loop back to both types equally. In sum, according to traditional models, natural selection is insensitive to a vast class of variation among individuals. It is incapable of selecting for any trait whose benefit arises from an indirect effect. This is the essence of the group-selection problem, applied to interspecific relationships. Although stated fairly simply here, the conclusion holds for any model of community coevolution in which each species/type is characterized by a single density (Levins 1974, 1975; Levin and Udovic 1977; Roughgarden 1976, 1977, 1978). For instance, Roughgarden (1976, p. 399–400), in a more formal community coevolution model, reaches identical conclusions in his "case of the nasty competitor."

Now we are in a position to reconsider the analogy between human and biological communities. They are similar in their complex webs of relationships. However, in human communities both direct and indirect effects are considered advantageous whenever their benefit exceeds the cost. Because indirect effects predominate in terms of effect on personal welfare, the individual usually has no choice but to act in the interest of the community. In traditional models of biological communities, the concept of a "community reaction" has no selective weight. Biological communities thus lack the cohesive force that binds human communities into a cooperating unit. Without this cohesion, there is no necessary relation between individual and community interest—and, as it is the former that is maximized, a community may be expected to be a somewhat disharmonious assemblage of species, exactly as many traditional evolutionary biologists envision it.

The Evolution of Indirect Effects in Structured Demes

Is natural selection's insensitivity to an individual's effect on its community truly a feature of nature, or an artifact of traditional models? It can be shown to disappear when the effect of deme structure is included.

The group-selection concept can be applied to community coevolution by simply considering the biotic community as a part of the environment of the species in question. A group-advantageous trait then consists of any modification of the surrounding community that increases the productivity of the trait group. The trait group is defined as before: any population that is homogeneous with respect to ecological interactions. In this case, the trait group corresponds to the set of conspecifics that benefits from the indirect effects of an individual *A*-type's activities. There is abundant evidence that biological communities are "textured" (Root 1975) such that indirect effects are often confined to a small area around the individuals that cause the effects (Wiens 1976, Levin 1976, Whittaker and Levin 1977).

Figure 4.1 shows a simple model of indirect effects. The *A*-type performs some action on the community, at a cost (c) to itself. The community reacts in such a way that feeds back positively to the trait group at large, including both the type that caused it and types that did not. As an example, earthworms are well known to stimu-

FIGURE 4.1 Model of indirect effects. The *A*-type stimulates the community, at a cost (c) to itself. The community's response results in a gain (g) for the *A*-type, but also for the *B*-type.

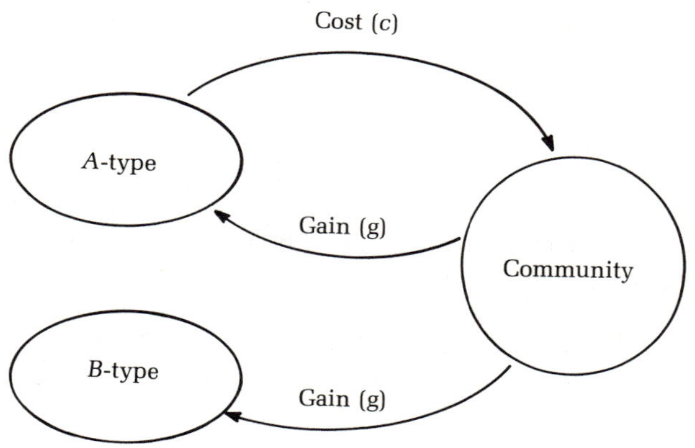

late plant growth through a variety of pathways, which eventually feed back as increased worm food. Yet the improved resource is available to all worms (as well as other species) in the trait group, not just the individuals that caused it.

If the reaction of the community is a linear function of the number of individuals that cause the reaction, then the per capita fitnesses of the two types are

$$f_A = Npg - c, \qquad f_B = Npg \tag{4.5}$$

Notice that this is identical to the linear model of Chapter 2 where d, the effect of an individual on itself, equals $(c + g)$. In unstructured demes, the A-type is selected for only if

$$\begin{aligned} f_A &> f_B \\ Npg - c &> Npg \\ 0 &> c \end{aligned} \tag{4.6}$$

The A-type therefore cannot be selected if the cost is to remain a cost. (If $c = 0$, natural selection cannot discriminate between the two types. This latter case corresponds to the matrix example, which didn't consider costs.)

In structured demes, trait groups with a high frequency of A-types enjoy a more highly modified community. The global fitnesses are

$$\begin{aligned} F_A &> F_B \\ Np_A g - c &> Np_B g \\ Ng(p_A - p_B) &> c \end{aligned} \tag{4.7}$$

and given a binomial trait-group variation,

$$p_A = p + pq/Np, \quad p_B = p - pq/Nq$$

$$\begin{aligned} Ng(p/N + q/N) &> c \\ g &> c \end{aligned} \tag{4.8}$$

We may now state the general conclusion: In traditional models a trait can be selected only if it provides a direct advantage to its bearer. Natural selection is insensitive to any trait whose advantage accrues through indirect effects. However, in structured demes with binomial trait-group variation, natural selection does not discriminate between direct and indirect effects. The trait is selected whenever its net effect on fitness, both direct and indirect, is positive for the individual manifesting the trait.

I believe that the entire process of evolution on the community level depends upon this conclusion. It exposes a vast class of variation among organisms to natural selection that have been hidden from traditional models. Structured-deme theory predicts that populations routinely evolve to stimulate or discourage other populations upon whom their fitness depends. As such, over evolutionary time an organism's fitness is largely a reflection of its own effect on the community, the reaction of the community to that organism's presence. If this reaction is sufficiently strong, only organisms with a positive effect on their community persist. The paradox between individual and community fitness is resolved, and the analogy between human and biological communities is complete.

A Clarification

Having gained this result, it is necessary to go over the same ground with a little more care. While the irrelevance of indirect effects is a necessary corollary of the individual-selection model, certain cases, such as mutualism, are readily accepted by traditional theory. Furthermore, human communities don't require structured demes for indirect effects to operate, and whatever powers indirect effects in human communities might also operate to some extent in nature. Even assuming that indirect effects are necessary for community evolution, what exactly is necessary for the evolution of indirect effects?

There are two pathways for the evolution of indirect effects, as there are for the evolution of altruism. In human communities, indirect effects are caused by behavioral structuring in the form of individual recognition, which makes it possible to channel the indirect effects of traits back to the individuals that caused them. Trivers (1971) realized that a form of "reciprocal altruism" could evolve through the individual recognition of altruists, and the concept has since been broadened to include all forms of indirect effects mediated by behavioral structuring, such as barter (e.g., E. O. Wilson 1975).

Individual recognition is certainly possible among many animals, in which case indirect effects may evolve without the mechanism of structured demes. For those organisms whose traits cannot be directed by individual recognition, the theory of structured demes provides an alternative pathway. (Trivers 1971 also verbally considered a form of spatial structuring similar to structured demes. See West Eberhard 1975 for a good summary.) The one class of traits accepted by traditional theory whose benefit arises from indirect effects is mutualism. A damselfish that feeds its anemone is obviously improving its own protection. But notice that mutualism involves a one-to-one interaction between individuals. The damselfish's behavior is directed back to only itself. In other words, it corresponds to a trait group of one.

Traditional theory encounters difficulty with what may be termed "population mutualism," in which the size of the trait group is greater than one and indirect effects must be shared with neighbors. This is probably the most common situation, and many examples of it will be provided in Chapter 5. However, it is not necessary to argue the relative contributions of the various pathways. The important point is that all of them together make it possible that the adaptive evolution of indirect effects is a general occurrence in nature, which has not been adequately considered by traditional treatments of community structure.

Some Examples

As an illustration of the relationship between individual and community fitness, consider the following imaginary example: Earthworms are well known to improve plant growth in a variety of ways. The passage of soil through their guts dramatically increases its nutrient value. They also improve the structure of the soil, aerating it and turning it over. Finally, they actually seem to secrete plant-growth substances into the soil (e.g., Witkamp 1971, Spedding 1971).

Because they are so beneficial to plants, and because plants form the foundation of any natural ecosystem, it is probably not too great an exaggeration to say that earthworms have a positive effect on nearly every member of their community. In other words, considering both direct and indirect effects:

$$\frac{\partial N_i}{\partial N_E} > 0$$

where N_E = the density of earthworms and N_i = the density of any other species in the community. The reverse, however, is probably not true, i.e.,

$$\frac{\partial N_E}{\partial N_i} = \text{variable}$$

One can well imagine that the ease with which the earthworm operates depends largely on the community surrounding it. Plant detritus may vary in the ease with which it is ingested. Root systems and their effects on the soil may vary the energy required to make burrows. Plant secretions may stimulate or inhibit.

Pathogenic bacteria may be present, which in turn are inhibited or stimulated by other components of the soil fauna, either directly or through changing the soil chemistry or microclimate.

Consider two types of plants (A and B) that are beneficial and detrimental to earthworm activity, respectively. Let per capita fitnesses equal

$$\frac{N_{A,t+1}}{N_{A,t}} = \frac{N_{B,t+1}}{B_{B,t}} = 1 + M_{A,B}\left[(N_{E,t}/(L + N_{E,t}))\,K_{A,B} - N_{A,t} - N_{B,t}\right] \quad (4.9)$$

$$\frac{N_{E,t+1}}{N_{E,t}} = 1 + M_E\left[(N_{A,t}/N_{A,t} + N_{B,t}))K_E - N_{E,t}\right] \quad (4.10)$$

These are modifications of the logistic equation, in which the carrying capacity of the plants ($K_{A,B}$) depends asymptotically on worm activity, while the carrying capacity of the worm (K_E) depends on the relative proportions of A and B. The constants M and K represent the rate of increase and the maximum carrying capacity. The constant L governs the rate at which the plant-carrying capacities asymptote. If I is the amount of time spent within the trait group (usually a single generation), then the time interval $(t + 1)$ is intended to be short relative to I. The short time interval is necessary to allow the feedback of indirect effects within the trait group. If the reader objects to changes in density during a single generation, the terms may be redefined as changes in the growth of individuals, which at reproduction translate into density changes.

As an individual selection model, this one is sufficient as it stands; Figure 4.3(a) gives the results for one set of parameter values (provided in the legend). The plants retain their starting proportions. Natural selection cannot discriminate between them because, although they vary in their effects on the earthworm, these effects feed back to both plant types equally.

Structured demes are modelled by computer simulation as follows. If the starting densities for the unstructured demes are N_A, N_B, N_E, a random-number generator is used to establish a number (T) of trait groups. The mean density for the trait groups is N_A, N_B, N_E, and the variance in density among trait groups is VN_A, VN_B VN_E ($V = 1$ in all examples). The covariance in density between species/types is zero. Equations (4.9) and (4.10) are then iterated I times separately for each trait group. This step corresponds to interactions within the trait group. Then the species densities are summed for all trait groups and divided by T to yield new mean densities per trait group. This step corresponds to the mixing phase. Finally, the process is repeated a number (Q) of times, or "trait-group generations." A flow chart of the simulation is provided in Figure 4.2.

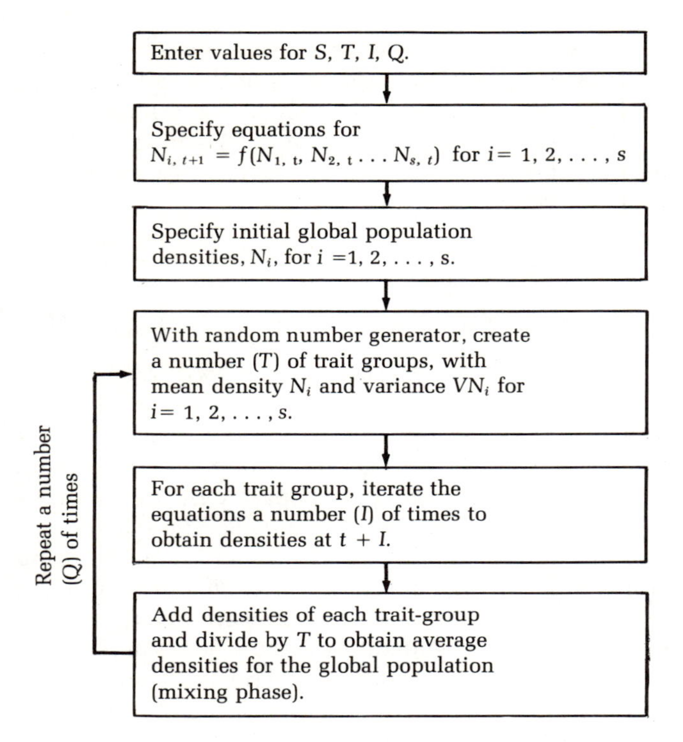

FIGURE 4.2 Flow chart of simulation model of structured demes.

The results for the structured deme simulation is shown in Figure 4.3(b). Due to the effect of deme structure, the plants differentially experience their effects on their community. Those trait groups with the highest frequency of plant *A* are superior to others. Plant *A* is selected over plant *B*, and the earthworm's carrying capacity is increased.

It is of interest to inspect some variations on this theme. Suppose, for instance, that the earthworm had a negative effect on the plants. This can be modelled by changing equation (4.9) to

$$\frac{N_{A,t+1}}{N_{A,t}} = \frac{N_{B,t+1}}{N_{B,t}} = 1 + M_{A,B} \left[(1 - N_{E,t}/(L + N_{E,t}))K_{A,B} - N_{A,t} - N_{B,t}\right] \quad \textbf{(4.11)}$$

Here the plant-carrying capacity decreases asymptotically with worm density. The results of the computer simulation for structured demes are shown in Figure 4.4(a). Now plant *B*, with its negative effect on earthworms, is favored by selection, with both plant *A* and the worm becoming extinct.

It follows that if an optimal density exists as far as the plants are concerned, the proper proportions of plants A and B will evolve to produce it. This may be seen by changing equation (4.9) to

$$\frac{N_{A,t+1}}{N_{A,t}} = \frac{N_{B,t+1}}{N_{B,t}} = 1 + M_{A,B}\left[(N_{E,t})(K_E - N_{E,t}) / (0.5K_E)^2)K_{A,B} - N_{A,t} - N_{B,t}\right] \quad \textbf{(4.12)}$$

FIGURE 4.3 (a) Evolution in unstructured demes. The plants retain the density ratio at which they started. (b) In structured demes, however, plant A is selected for, with resulting increases in worm density. Plant B eventually becomes extinct. (For this run, $M_{AB} = M_E = 0.01$, $K_{AB} = 20$, $K_E = 25$, $L = 15$, $V = 1$, $I = 10$.)

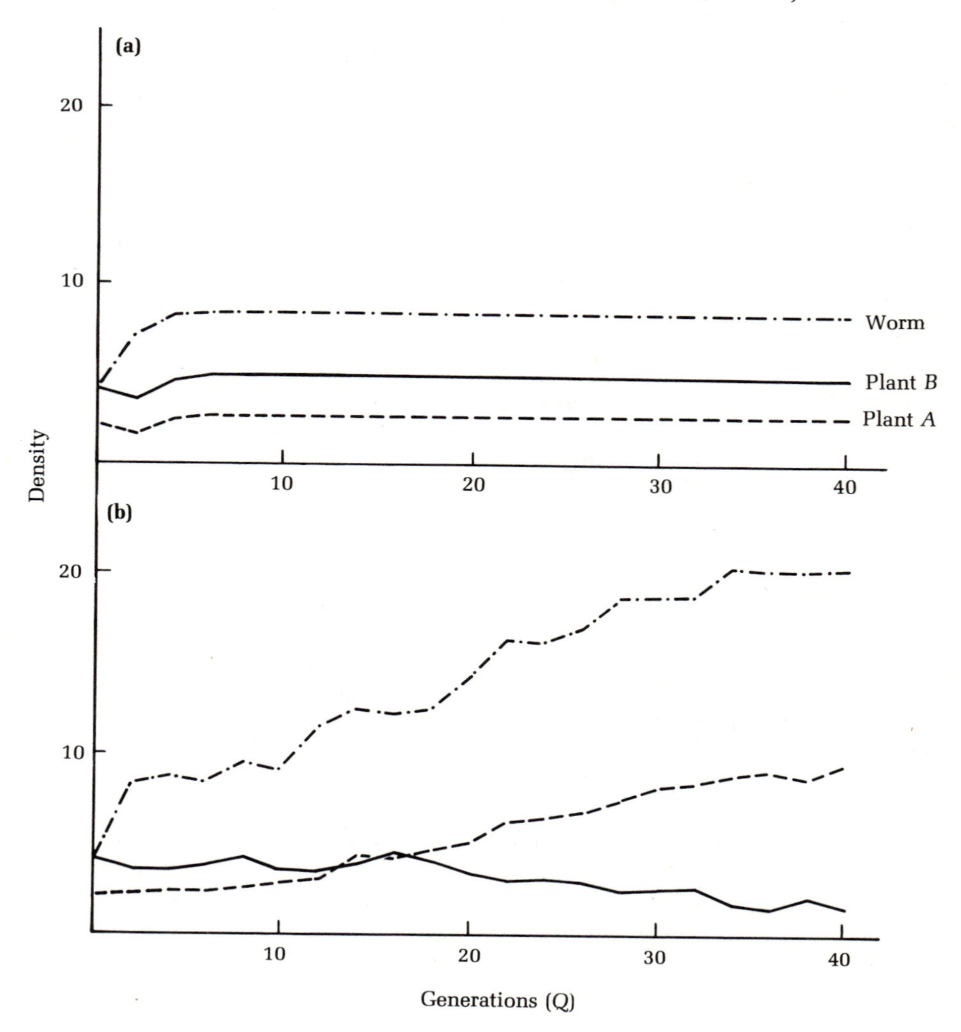

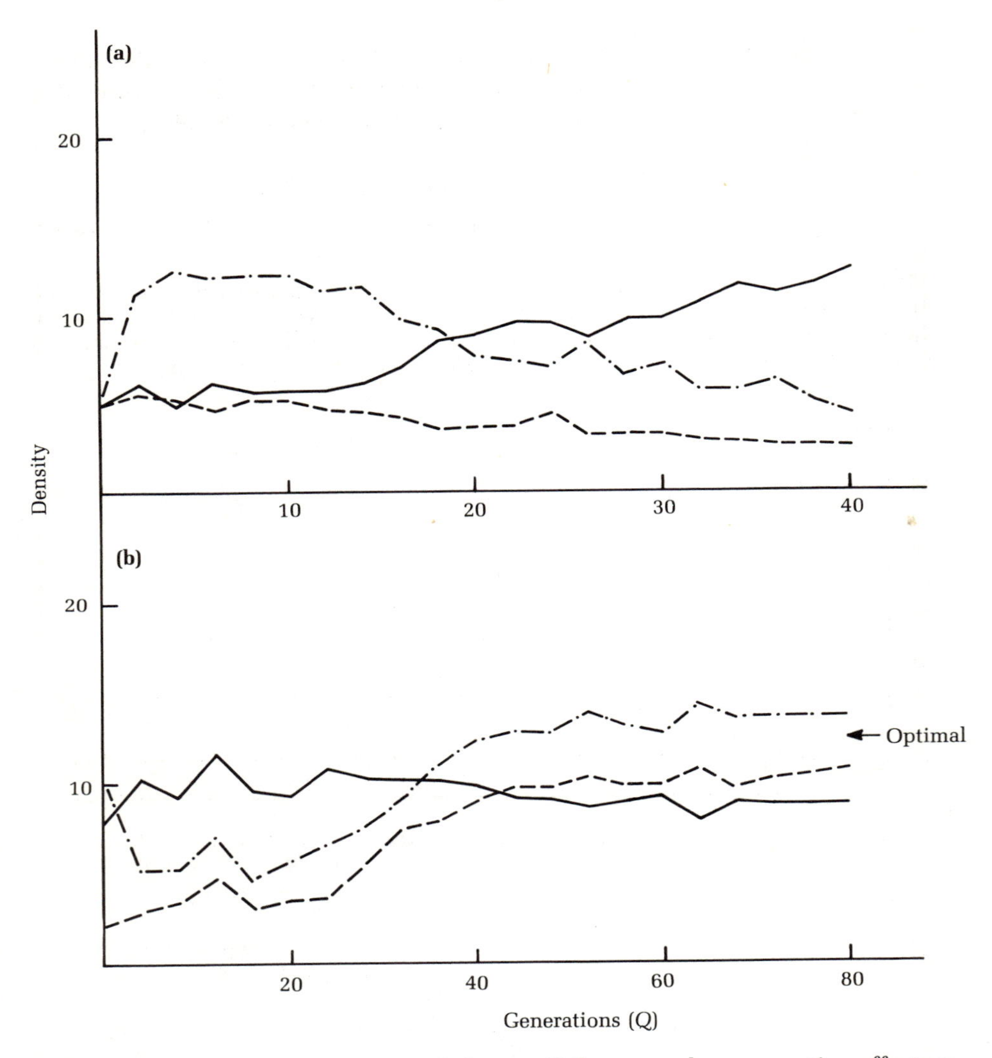

FIGURE 4.4 Evolution in structured demes. If the worm has a negative effect on the plants (a), both worm and plant A are driven to extinction. If an optimal worm density exists for the plants (b), the proper proportions of plants A and B evolve to produce it. Parameter values are the same as for Figure 4.3.

Now plant-carrying capacities are maximized at a worm density of $0.5K_E$. The results for structured demes are shown in Figure 4.4(b).

As a final permutation, return to the situation in which earthworms are always beneficial to plants (equation 4.9), and assume that instead of directly benefiting the earthworm, plant A produces a microhabitat that stimulates bacteria that inhibit a fungus that is detrimental to the earthworms. In other words, the indirect effect consists of three links instead of one. This situation can be modeled by the following series of equations:

$$\frac{N_{A,t+1}}{N_{A,t}} = \frac{N_{B,t+1}}{N_{B,t}} = 1 + M_{A,B} \left[(N_{E,t}/L + N_{E,t})\ K_{A,B} - N_A(t) - N_B(t)\right] \quad \textbf{(4.13)}$$

$$\frac{N_{c,t+1}}{N_{c,t}} = 1 + M_c\left[(N_{A,t}/(N_{A,t} + N_{B,t}))K_c - N_{c,t}\right] \quad \textbf{(4.14)}$$

$$\frac{N_{D,t+1}}{N_{D,t}} = 1 + M_D\left[(1 - N_{c,t}/K_c)K_D - N_{D,t}\right] \quad \textbf{(4.15)}$$

$$\frac{N_{E,t+1}}{N_{E,t}} = 1 + M_E\left[(1 - N_{D,t}/K_D)K_E - N_{E,t}\right] \quad \textbf{(4.16)}$$

The results are shown in Figure 4.5. Plant *A* is selected for, with corresponding changes in all the other members of the community. Notice that with an increased number of links in the indirect effects, more iterations of the equations per generation are needed for selection to occur. (Whether it takes longer in terms of real time is discussed below.)

To summarize, in this model the earthworm's existence depends on the function it performs in its community. Before proceeding, it is desirable to inspect the assumptions behind the model closely, to see exactly what is responsible for its main conclusion.

First, it is of course necessary for structured demes to exist and for the benefit of the indirect effect to exceed the cost (given random trait-group variation). If trait-group variation is greater than random, then the cost may exceed the benefit, and the trait becomes strongly altruistic.

Second, the indirect effects must loop back within a relatively short period, possibly within the lifetime of the individuals that cause them (see p. 41). Whether this is a serious limitation remains to be determined. Certainly some traits could take generations to manifest themselves, but others can occur on a very short time scale. The example just given, in which a change in microclimate stimulates bacteria that inhibit a fungus, could occur within days. Other examples of indirect effects whose benefits occur within the lifetime of the individuals causing them are given in Chapter 5.

Third, in the examples used, the plants depend completely on the earthworm and the earthworm depends completely on plant *A* ($K = 0$ without the essential species). Had this not been true for the

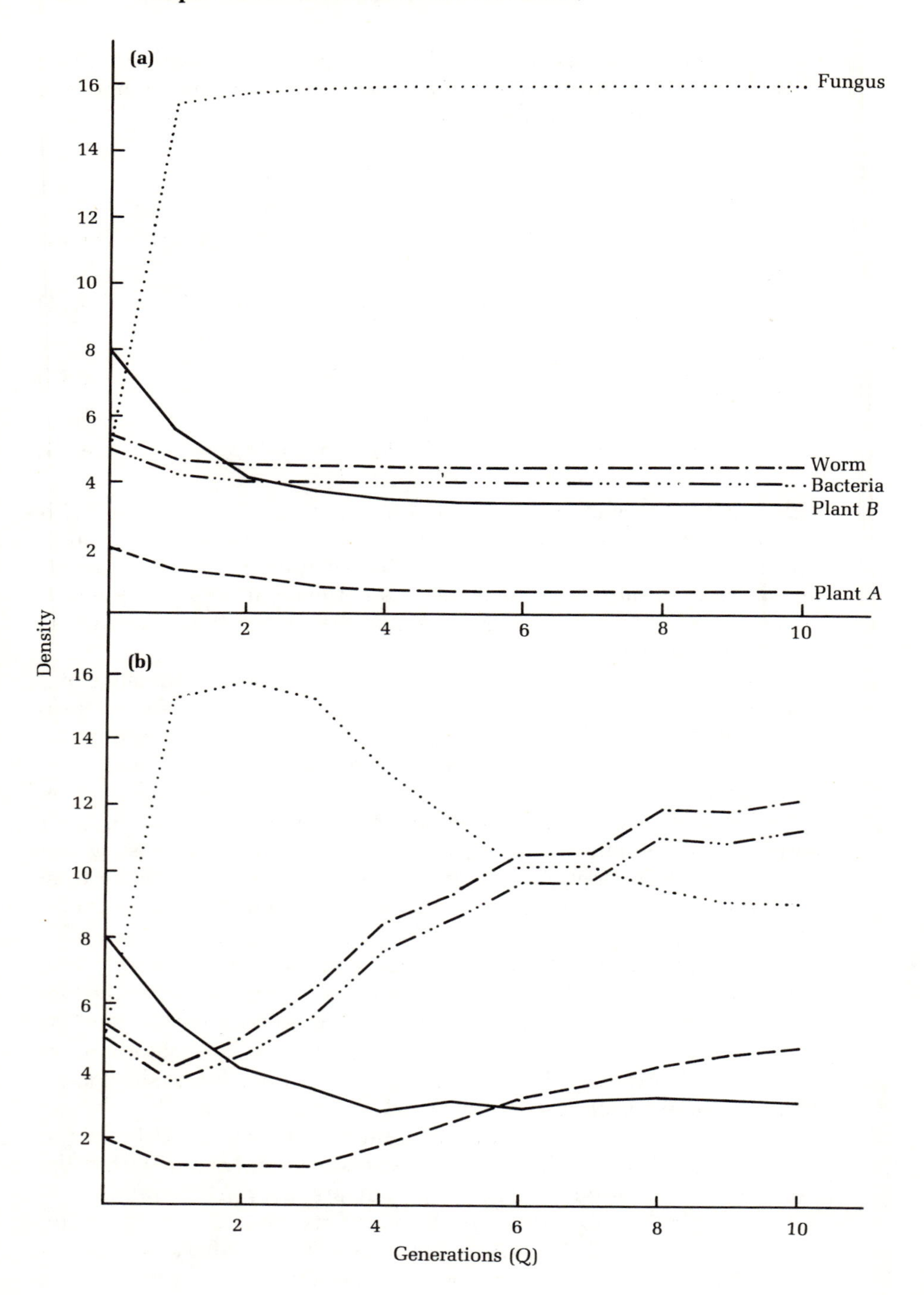
(a)
Fungus
Worm
Bacteria
Plant B
Plant A
(b)
Density
Generations (Q)
16
14
12
10
8
6
4
2
2
4
6
8
10

example in which the earthworm had a negative effect, the worm would have become less common instead of extinct, but plant *A* would still have been replaced by plant *B*. While the complete control assumption is unrealistic for this example, it is likely that multispecies communities do have virtually complete control over their members, and the only way to simulate complete control in a simple model is to increase the amount of control per species. Furthermore, total control is not necessary for other aspects of community evolution, as discussed below.

These are the main ingredients of the model; while some additional simplifying assumptions are made, they are not critical. For instance, the plants are identical except for their effect on the earthworm. While this assumption is unrealistic, it is nothing more than the "all other things equal" clause used in all evolutionary arguments. A more realistic and equivalent assumption is that the genetic determinants of the trait in question vary independently of other traits, or are the dominant of several pleiotropic traits in terms of effect on fitness (in which case, the negative effects of the other traits are included in the cost).

In his classic defense of individual selection, G. C. Williams (1966) also used the example of the earthworm.

> *On the other hand, suppose we did find some features of the feeding activities of earthworms that were inexplicable as trophic adaptations, but were exactly what we should expect from a system designed for soil improvement. We would then be forced to recognize the system as a soil modification mechanism, a conclusion that implies a quite different level of adaptive organization from that implied by the nutritional function. As a digestive system, the gut of a worm plays a role in the adaptive organization of that worm and nothing else, but as a soil modification system it would play a role in the adaptive organization of the whole community. This, as I will argue at length in later chapters, is a reason for rejecting soil-improvement as a purpose of the worms' activities if it is possible to do so. . . .*

← **FIGURE 4.5** Evolution in unstructured (a) and structured (b) demes. Plant *A* increases the fitness of bacteria, that decrease the fitness of a fungus, that decrease the fitness of the worm. In structured demes, plant *A* is selected, with corresponding changes in the community. For this run, all *M*'s = 0.01, $K_{AB} = K_C = K_D = 20$, $K_E = 22$, all *L*'s = 15, $V = 1$, and $I = 30$. The increased number of links in the indirect effect necessitates an increased number of iterations (I) per generation.

If the theory presented here is valid, this view must be modified in several ways. First, even if the earthworm's activities are devoted entirely to maximizing feeding efficiency, it may still be constrained to behave in the interests of the community. The examples presented here show that regardless of the worm's internal motivation, it can truly be said to exist for its role in the adaptive organization of the whole community.

Second, if acting as a soil modification mechanism improves the environment through indirect effects for the earthworm (a kind of farming), it can evolve in a direction that really is inexplicable in terms of direct trophic adaptation. In the examples, the earthworm was treated as a nonevolving entity, the selective force to which the plants responded. It would have been just as simple to construct models of earthworms evolving to stimulate plants, and in nature one would, of course, expect both.

The interplay between the evolution of a species and the evolution of its community is very fundamental and perhaps can be made clearer with another imaginary example. Boothe and Knauer (1972) found that the crab *Pugettia producta* (Randall) concentrates certain trace elements in its feces. This is likely to mean very little directly to the crab, but it means more to a large set of other species in the community that require those nutrients for growth. Among these species one might expect variation in their effect on the crab. The variation can act along multiple pathways, including the inhibition of species that themselves inhibit the crab. Through the process of selection in structured demes just described, an assemblage evolves that enhances the environment for the crab.

Now the crab is embedded in a community that has a definite relationship with it. It too will evolve in a direction that is beneficial to itself through indirect effects. But how? One way is by increasing the concentration of nutrients in its feces. In this way the community directs the evolution of the species, in much the same way as human communities direct the professions of its members. Notice that it is not necessary for the community to have total control over its members, only enough to make it advantageous to evolve in the proper directions.

Natural Selection on the Level of Communities?

I have attempted to construct a model of multi-species coevolution in which spatial variation in the genetic composition of populations is taken into account. The actual mechanism is no different than described for group selection in Chapter 2, the only difference being that the community is considered as part of the environment of each species. Is there any justification in calling the process a form of natural selection on the level of communities? Or rather, since the question is largely semantic, is there any utility in making such a distinction?

Any mechanistic model of community selection would have to include the following ingredients: (1) a population of communities that (2) vary in a heritable fashion (the heritability condition is satisfied if differences in species fitness are caused by underlying differences in species composition), and (3) a criterion of selection that governs the differential productivity of the variants. Natural selection would then shift the composition of the community in the direction of increased productivity.

It is clear that the concept of structured demes lends itself as well to small-scale variation in species composition as to the genetic composition of single-species populations. The key distinction between group selection and community selection, therefore, revolves around the form of variation among competing types. If the variation is *intra*specific—that is, if the two types in the preceding models represent genotypes within a single species—then natural selection changes the genetic composition of the population (group selection). If the variation is *inter*specific, such that the two types in the preceding models represent different species, then natural selection changes the species composition (community selection). Of course, the models themselves make no distinction as to whether the types are of the same or different species, and a consideration of diploidy does not alter the main conclusions. In this sense, group and community selection involve exactly the same process, and the distinction between intraspecific and interspecific variation is trivial.

In my opinion, the coevolutionary process described in this chapter can be usefully thought of as community selection if it actually leads to well-organized assemblages, mutualistic networks in which most species can be demonstrated to have a net positive effect on their community. The prospects for such superorganisms are considered empirically in Chapter 5 and theoretically in Chapter 6.

5 The Evolution of Community Function

In Chapter 4 I attempted to show that organisms practice a form of biological control, modifying the community to their own advantage whenever possible. To proceed further, it is necessary to ask some very naïve questions. What kinds of control can organisms exercise over their community? What kinds of relationships exist among organisms that can evolve in the direction of enhancement or inhibition?

If one attempted to answer these questions from the literature on communities—especially the theoretical literature—he or she would conclude that organisms enter the lives of others mainly as predators, competitors, or prey. (See Colwell and Fuentes 1975 for a refreshing exception.) Again and again, treatments of biological communities pay passing tribute to "complex interrelationships among species" and then lapse into food webs and competition models as if nothing else existed. As a recent example, Connell (1975) states, "If a local assemblage of organisms is to be regarded as a community with some degree of organization or structure, then it is in the interactions between organisms that we must look to provide the structure. Two different interactions provide most of the organization: competition and predation."

One way to test the generality of this framework is to look for examples that do not fit into it. Table 5.1 presents (by no means exhaustively) a list of "nontrophic" interactions—relationships among organisms that very much affect fitness but not in the form of competition or predation. Some interesting examples from Table 5.1 are:

1. Neill (1975) studied a laboratory microcosm consisting of zooplankton, algae, and a benthic amphipod.

The amphipod had a major stablizing effect on its community by its modification of the physical habitat. The vast majority of algae grew within old amphipod fecal pellets, and only after the fecal pellets disintegrated with age were the algae released into the water column where they were grazed upon by zooplankton. The amphipods thus created a refuge for algae that prevented overgrazing by zooplankton.

TABLE 5.1 Some examples of nontrophic interactions: relationships among organisms that affect fitness but not in the form of competition or predation.

Description	*Reference*
Grass shrimp restructures fine texture of detritus, creating cavities that promote algal attachment and growth.	Welsh 1975
Benthic animals bind sediments, increasing suitability for other species.	Rhoads and Young 1970, 1971
Ants restructure soil, increasing suitability for certain plant species.	King 1977
Buffalo habitat dependent upon activities of hippo.	Field 1970, Eltringham 1974
Hippo-wallowing creates water holes that permit more extensive range utilization by other animals.	Field 1970
By removing sea urchins, sea otters allow the existence of kelp beds, which in turn leads to major faunal changes, including changes in neighboring terrestrial ecosystems.	Estes and Palmisano 1974
Dung beetles bury dung, making it inaccessible to Diptera, thereby influencing the transmission of disease to livestock.	Hughes et al. 1974, Waterhouse 1974
Many dung insects depend upon the tunnels of beetles to forage.	Valiela 1974
Colonization of rock areas by coral is dependent upon removal of algae by echinoids.	Dart 1972
E. coli vulnerability to phage attack depends upon presence and nature of sediment, which can be influenced by a variety of biotic factors.	Roper and Marshall 1974
Earthworms promote turnover, structuring of soil.	Witkamp 1971
Litter characteristics alter searching efficiency of parasitoids.	Price 1970
Water lettuce (*Pistia stratiotes* L.) influences acidity, stratification, oxygenation, solar radiation, and temperature of the water below it.	Attionu 1976

(continued)

TABLE 5.1 (Continued)

Description	*Reference*
Filamentous bacterium *Beggiatoa* reduces hydrogen sulfide levels in soils, increasing the fitness of rice plants. Rice plants also benefit *Beggiatoa* by an unknown mechanism.	Joshi and Hollis 1977
Floating vascular plants increase water temperature in early spring.	Dale et al. 1976
"Understory" species depend upon "canopy" species for existence in intertidal algal community.	Dayton 1975
Mountain laurel conserves phosphorus in system by its evergreen nature.	Thomas and Grigal 1976
Crab concentrates trace minerals in feces.	Boothe and Knauer 1972
Digestive activities of carp are a major contributor to the nutrient loading of lakes.	Lamarra 1975
Soil structure and chemistry is determined by plants.	Crocker and Major 1955, Zinke 1962, Pahlsson 1974
Dung beetles speed and redirect the cycling of nutrients from dung to plants.	Hughes et al. 1974, Waterhouse 1974
Algal exudates have chelation properties, increasing the availability of trace elements.	Johnston 1964
Amphipod feces provide refuge for algae, making them inaccessible for planktonic grazers and stabilizing the community.	Neill 1975
Algal mat provides refuge for fly.	Collins 1975, Collins et al. 1976
Rodent burrows improve water retention of soil, provide habitat and refuge for invertebrates and other vertebrates.	Naumov 1975, Golley et al. 1975, Batzli 1975
Aquatic macrophytes serve as refuge for benthic animals during unfavorable (e.g., dry) periods when the mud itself is uninhabitable.	McLachlan 1975
Coral provides refuge for a variety of animals.	Glynn 1976
Crustaceans protect coral from Echinoid predation.	Glynn 1976
Host-finding behavior of flea beetles is impaired by the chemical stimuli given off by neighboring nonhost plants.	Tahvanainen and Root 1972
Plants form "defense guilds," protecting each other from predation.	Atsalt and O'Dowd 1976, Feeny 1975, Root 1975.
Parasites modify rate of predation on intermediate host.	Bethel and Holmes 1977
Plant growth regulators are produced by microorganisms.	Katznelson and Cole 1965

TABLE 5.1 (Continued)

Description	*Reference*
Greatest mortality of parasites occurs during time spent between hosts, and can be influenced by a variety of biotic factors.	Kennedy 1974
Ants have major effect on plant utilization by other species. Effects differ radically with the species of ant.	Bentley 1976; Leston 1970, 1971
Fish ectoparasites are decreased by activities of cleaning fish.	Losey 1974
Phoresy is the use of an organism for transport.	(reviewed below)
The microbes *Proteus vulgaris* and *Bacillus polymyxa* each synthesize a vitamin which the other needs but cannot synthesize itself.	Yeoh, Bungay, and Krieg 1968
A strain of *Arthrobacter* and a strain of *Streptomyces* are both necessary to break down the pesticide "*Diazinon*."	Gunner and Zuckerman 1968
Germination and growth of fungus are dependent upon a variety of biotically determined factors.	Lockwood 1977
A variety of interactions are mediated by chemical interactions in aquatic environments.	Lucas 1961, Fogg 1966, 1971, 1975
Recruitment and life history patterns of limpets are dependent upon community composition.	Lewis and Bowman 1975
Aquatic plants inhibit mosquito populations through a variety of pathways.	Angerelli and Beirne 1974
Grasshoppers promote growth of grass, beyond the mechanical effects of clipping. There are possibly growth-promoting substances in saliva.	Dyer and Bokhari 1976

2. Welsh (1975) studied the role of the grass shrimp (*Palaemonetes pugio*) in a tidal-marsh ecosystem. This animal apparently increases productivity by altering the microstructure of the detritus it feeds upon. Scanning electron microscope studies showed that its feeding activities create tiny cavities which become heavily colonized by diatoms. In addition, the grass shrimp had other profound effects on its community. The author concludes that "*Palaemonetes pugio*, while supporting its own trophic requirements, accelerated breakdown of detritus, preventing blockages or ac-

cumulations that might have occurred from pulses of emergent grass and macroalgal detritus in the embayment. This repackaging into feces, heterogeneous fragments, DOM, and shrimp biomass made detrital energy available at a variety of trophic levels, smoothing out organic pulses over time and space, and raising the efficiency of transfer to the food web."

3. Angerelli and Beirne (1974) studied the effects of ten species of aquatic plants on the survival and oviposition behavior of mosquitos. Three had no demonstrable effect and one acted directly as a predator (the bladderwort *Utricularia minor*). All the rest had a nontrophic effect. For instance, the alga *Chara globularis* produced a juvenile hormone-like substance. The duckweed *Lemna minor* also liberated an inhibitory substance into the water, and by covering the surface of the water, inhibited oviposition by adults and the respiration of larvae. *Callitriche palustris* supported hydra along its stems, which preyed upon mosquito larvae, and *Nymphaea tuberosa* served as an oviposition site for predatory beetles. Other plants are known to stimulate oviposition by mosquitos (Rapp and Emil 1965). It seems clear that the suitability of a habitat for mosquitos can depend largely on the composition of plants in the mosquitos' trait group.
4. Bethel and Holmes (1977) have produced evidence that acanthocephelan parasites can change the behavior of their intermediate hosts (amphipods), increasing the amphipods' vulnerability to predation by the parasite's definitive host (mostly waterfowl). In addition to fulfilling its own trophic requirements, the parasite is also modifying the rate at which a predator–prey interaction takes place.
5. Atsalt and O'Dowd (1976) have suggested that some plant species interact to form "plant defense guilds." In other words, several species, each employing a different strategy, are necessary to ward off herbivores. If true, then some plants enter the lives of others not only as competitors but also as protectors. (See also McNaughton 1978.)
6. Productivity in tundra ecosystems is often determined by the permafrost level, which in turn can be modified by lemming activity (reviewed by Batzli 1975).
7. Rhoads and Young (1970, 1971) found that *Molpadia*

volitica (Holothuroidea) reworks the bottom sediments of Cape Cod Bay, stabilizing it and increasing its suitability for other benthic organisms.

The relationship between trophic and nontrophic interactions can be explored using the standard Lotka-Volterra equation:

$$dN_1/dt = \frac{MN_1}{K}(K - N_1 - a_{12}N_2) - a_{13}N_1N_3 \qquad \textbf{(5.1)}$$

where N_1 is the density of the organism in question; N_2, the density of a competitor; and N_3, the density of a predator. K and M are the carrying capacity and rate of increase. a_{12} and a_{13} are the coefficients of competition and predation, respectively. In field experiments one often tests the importance of predators by removing them, and if predators exert an influence, then a response is duly recorded. Competitors can similarly be removed or augmented. By this method, it is shown that competitors and predators have a pervasive influence on organisms in nature. (Connell 1975 and Colwell and Fuentes 1975 review the evidence.)

The point is that while this approach to studying competition and predation is valid, it is valid for countless values of K, M, and a. Yet all of population biology's cherished constants—its rates, coefficients, and carrying capacities—are in fact variables, governed by the composition of the surrounding community in the form of nontrophic interactions. Baker and Cook (1974, p. 16) state: "The progression through the stages of the life cycle of a soil microorganism is determined at least as much by the associated microflora and abiotic environment as by the genes of the microorganisms." The same is probably true of organisms in general. The entire structure of biological communities, the flow of elements, and the rates of reaction are heavily influenced by its constituent species.

To summarize, traditional investigations of community structure begin by assuming that most of the structure is already there. They focus on relatively superficial forms of competition and predation, while ignoring the deep structure that determines the parameter values of the models in the first place.

The concept of nontrophic interactions is certainly not new. In fact, the whole subdiscipline of ecosystem structure and function is largely devoted to its study (Odum 1971). Why then has it remained so largely untouched by evolutionary theory? That is precisely the problem: Nontrophic interactions concern community structure and function, which according to individual selection models are not subject to adaptive evolution. One does not speak of the adaptive evolution of nutrient cycling, stability, or the relation between an animal and its refuge, except within the rather limited sphere of individual mutualism. These interactions among organisms are thought to be the by-products of the evolutionary process.

Because nontrophic interactions so largely manifest themselves through indirect effects, the structured-deme model predicts that a vast array of relationships—at present ignored by most evolutionary biologists—are subject to natural selection. If true, then what are the implications for the study of biological communities? In my opinion, a functional approach becomes necessary. In other words, to truly understand the presence of many species in their communities, it will be necessary to discover their roles; their activities that feed back positively to the rest of the community.

In this chapter I hope to demonstrate the necessity of a functional approach for one type of natural community, using nonimaginary examples. In my opinion, the system I have chosen constitutes a particularly good field test of the structured-deme model and its predictions, but the same processes apply to a broad array of communities. The general tendency of species to evolve dependencies on other members of their community is discussed in a brief section on specialization. The chapter concludes by speculating on a possible method of biological control.

Phoresy

The term "*phoresy*" refers to the use of one animal by another for transport. Hardly an exotic interaction, it has evolved hundreds of times in as many taxa. (See reviews by Clausen 1976, Lindquist 1975, MacNulty 1971, Farish and Axtell 1971, Disney 1970.) Mites ride from flower to flower on the bills of hummingbirds (Colwell 1973), blackfly larvae attach to prawns and mayflies (Disney 1971, 1970), gastropods affix to hermit crabs (Hendler and Franz 1971); and given the proper circumstances, almost any beetle picked at random will be crowded with mites, gathered under the elytra, or attached to the ventral surface. Although usually much lower, densities can exceed 1000 mites per beetle (personal observation). The proper circumstances are habits and habitats of the carrier that guarantee the next generation of passengers (who develop in proximity to, but apart from their host) a reasonable probability of getting another ride.

Costa's (1969) work on the association between mesostigmatic mites and Coprid beetles provides a model example of phoresy. Both male and female beetles bury dung in the ground, which is eventually shaped into balls within which eggs are deposited. Eight species of mesostigmatic mites are associated with *Copris hispanus* (the species most intensively studied by Costa) with varying degrees of specificity. The mites abandon the beetle upon landing, and reproduce on the dung pad. Mites not only inhabit the dung but are also found in large numbers within the dung balls, along with the developing beetle larvae. The life cycle of the mites is highly synchronized to the beetle; although the mites' developmental time is invariably shorter, this is compensated for by a resting stage. Finally, the mites climb onto the beetle as it emerges from the pupa.

Aside from being fascinating in their own right, phoretic relationships allow us to make a prediction that distinguishes between individual selection and structured demes by asking the question, "What is the relationship between the passenger and the carrier during the nonphoretic stage, when both are developing side by side in the same habitat?" Because the beetle is essential to the mite for dispersal, one might expect the mite to evolve along pathways that enhance the survival of its carriers. But while the mutant mite that improves the fitness of the beetle does indeed improve its own chances for dispersal, it also improves the chances for every other mite in its trait group. This is an indirect effect and cannot be selected by the evolutionary forces operating within a single trait group. The structured-deme model thus predicts that passengers will routinely evolve a positive relationship with their carriers during the nonphoretic phase of the life cycle, while individual selection models predict they will not.

But this is not the only prediction; in fact, the beauty of phoretic systems is in their diversity. One can vary almost every important parameter in the mathematical model, and find corresponding phoretic systems in nature. For instance, phoretic systems exist in a continuum from individual mutualism, in which one mite associates with one carrier (e.g., Skaife 1952) extending to enormous trait groups with hundreds of unrelated passengers and dozens of unrelated carriers (e.g., dung beetle communities or bark beetle galleries). The effect of trait-group size on the evolution of indirect effects can thus be tested experimentally by seeing if the tendency to form positive relationships decreases with increasing trait-group size. Similarly, one can find phoretic systems that correspond perfectly to discrete trait groups (e.g., the Coprid beetles just discussed) and those that approximate the continuous model (e.g., the mites phoretic on sciarid flies in compost: Binns 1972, 1973). A continuum can be formed from a single mite generation between dispersal to many mite generations between dispersal (e.g., Lindquist 1969). Another continuum can be formed from simple two-species associations (e.g., Springett 1968) to highly diverse associations, such as the southern pine bark beetle (*Dendroctonus frontalis* Zimmermann) community that consists of more than 90 species of mites and 90 species of insects (Moser and Roton 1971; Moser, Thatcher, and Pickard 1971; for other complex phoretic systems, see Matthewman and Pielou 1971, Pielou and Verma 1968). Some carabid beetles even have as many as seven interconnected spaces beneath their elytra, with several species of mites, each morphologically adapted to occupy a different space (Regenfuss 1972, discussed by Lindquist 1975)—literally, a niche diversification!

Finally, one can find variation in the most important parameter of all, the degree to which the passenger depends upon the carrier for dispersal. Some passengers are completely dependent upon a single-host species, others upon many (Krantz and Mellott 1972). Others have alternative means of dispersal and are found not only in association with their carriers, but also at large in the soil or litter (e.g., Costa 1969). Moreover, some passengers are dependent upon a single species of carrier, but not upon the particular individuals with which they develop in the trait group (e.g., Clausen 1976). We will return to this important point later in this chapter.

Phoresy, therefore, represents an ideal system in which to investigate the structured-deme concept in nature. While not nearly enough is known to test the validity of the theory, we can get a slight indication of the kinds of relationships that exist between passengers and carriers, as well as the patterns that they form with respect to the continua outlined above.

By far the most thorough analysis of a phoretic relationship is due to Springett (1968) on the carrion beetles *Necrophorus humator* 01. and *N. investigator* Zett. and their mesostigmatic mite *Poecilochirus necrophori* Vitz. *Necrophorus* beetles have a social organization. Several pairs of beetles cooperatively bury a corpse, but eventually a single pair monopolizes it and lays all the eggs, the others dispersing (see also Milne and Milne 1976). The larvae feed on the corpse, but the adult female remains in the burial chamber and feeds them at the beginning of every instar, probably to reinoculate them with gut microflora lost during molting. Each beetle carries an average of 10 to 40 mites (depending on the season), which abandon their carrier upon landing on the corpse. Thus, the trait group in this case is composed of the mixed-mite fauna from all the beetles who cooperatively buried the corpse.

Springett (1968) investigated the relationship between the passenger and the carrier during the nonphoretic stage by setting up a three-species community composed of the beetle, the mite, and the beetle's chief competitor, *Calliphora* (Diptera) larvae. Communities were also created in which various members were absent. *Necrophorus* reproduced successfully either alone, with its mite, or in the complete three-species community. However, when only *Necrophorus* and *Calliphora* larvae were present, the beetle was completely unsuccessful at raising offspring. The pattern is explained by the fact that upon leaving their carrier, "the mites run rapidly over the corpse and, on finding a batch of *Calliphora* eggs, immediately stop, pierce the eggs with the chelicerae, and eat the contents. A corpse is normally clear of *Calliphora* eggs in less than four hours, and *Necrophorus* can reproduce without the presence of their chief competitors." Most interesting is the fact that even when the mite was provided fly eggs, it could not survive without the presence of *Necrophorus*. There is evidence that the adult female beetle feeds the mites on a regurgitated brown fluid from the foregut.

To summarize, the nonphoretic stages of *Poecilochirus* have a positive relationship with *Necrophorus* by eating the eggs of its competitor. It is of interest to try to imagine the set of roles *Poecilochirus* could have filled. Instead of eating fly eggs, it could have eaten nematodes, or beetle larvae, or the corpse itself. The fact that its choice of food happens also to be important to the survival of its carrier, and therefore indirectly to the mite population, may be coincidental, or it may be the dominating factor governing the evolution of the relationship. Had it been a single mite interacting with a single beetle, the answer would appear obvious, but the fact that it is many mites descended from perhaps ten or more unrelated parents creates theoretical problems for the individual selectionist—problems which possibly are solved by the theory of structured demes. One example may indeed be coincidental, but many examples gathered from diverse taxa of carrion beetles and mites (Borchers 1968), all showing positive indirect effects, would be a clear indication that group selection operates strongly at this level of trait-group size and variation.

Unfortunately, further examples do not yet exist, and the criteria for determining the effect of the passenger on its carrier are stringent. Springett was fortunate to have included *Calliphora* in his experiments. Had he not, he would probably have deduced a commensal relationship with a neutral effect of the mite on its carrier. Indeed, after a thorough study of a Carpenter bee and its mite in artificial nests, Skaife (1952) could find no effect of the mite on its carrier, even though it is a traditional mutualism: The female bee has a special abdominal pouch for the sole purpose of carrying mites and the pupa feeds the mites from secretions. The only way to be sure is to follow the fate of the carriers, with and without their passengers, in the field.

I am aware of only one other study that gives insight into a possible relationship of the passengers on the carriers. In a thorough behavioral study of a cercomegistid mite and its bark beetle associate, Kinn (1971) found two positive indirect effects. The mite preyed on the eggs and larvae of another mite species which in turn preyed on the eggs and larvae of its carrier; the mite also preyed on endoparasitic nematodes of its carrier. In particular, Kinn describes the mites clustering around the feces of the beetle, picking out the nematodes. Hence, we have a possible case of a passenger that feeds on the predators of its carrier, although the effect on barkbeetle fitness was not measured.

To proceed further with an examination of phoresy, it is necessary to examine the much larger literature on passengers that are predatory upon their carriers (the larger literature is due to the economic orientation of the field, and is not necessarily indicative of phoretic relationships in general). While predation may at times be beneficial to an overcrowded population or outweighed by other activities (for instance, Russo (1926, 1938) found that several species of mites preyed not only on bark beetle larvae but on their hymenopterous parasites as well), intense predation can probably be assumed to be detrimental to the carrier. If so, the structured-deme theory predicts an inverse relationship between the intensity of predation and the degree of dependence on the carrier. It is possible, therefore, to order phoretic relationships along a dependency gradient. For instance, many beetles of the family Meloidae lay their eggs in the vicinity of flowers. The first instar larva attaches to a solitary-nesting bee visiting the flower and is taken to the bee's nest, whereupon the beetle devours not only the bee larva but also its food provisions. Upon metamorphosing, the adult beetle flies away to repeat the cycle. Phoretic relationships such as these are fascinating because, while the beetle is dependent upon the bee for transport, it is not dependent upon the particular bee larva in its trait group. The indirect effect of its actions are negligible, and we would not expect a positive relationship to form between the passenger and the carrier, unless it were coincidentally beneficial to the former. Clausen (1976) exhaustively reviews the literature on this type of phoresy, and in many cases the passenger is a highly destructive predator.

A moderate degree of dependency is found in the mite-sciarid fly interaction investigated by Binns (1972) and the mite-housefly interaction studied by Farish and Axtell (1971). The habitats of both are subject to a relatively continuous traffic of egg-laying females, and the life cycles of the mites are not synchronized with their carriers. Indeed, *Macrocheles muscaedomisticae*, the mite that associates with the housefly, can develop from egg to adult in as little as three days (Wade and Rodriguez 1961). The mites from both studies feed on eggs and first instar larvae, offspring of their past carriers. However, their future carriers are already too large to be preyed upon, and the mites therefore suffer few indirect effects from their own predatory activity. Still, the situation deserves a much closer analysis. The mite studied by Binns (1972) is a very general predator, and there is also a second species of mite, phoretic on the same carrier, also a general predator but not on fly eggs (Binns 1973). Furthermore, a community consisting of two types of mite, one that preys on fly eggs and one that does not, can be shown to be a regulator of optimal fly density, in a manner similar to the plant-worm community described in Chapter 4.

At the opposite extreme of dependency are the mites of the genus *Iponemus*, which parasitize the eggs of bark beetles. Lindquist (1969b) has shown that 14 of 16 species have a monospecific relationship with their carrier. The female feeds upon a single egg, which is sufficient for both herself and her entire progeny, who do not themselves feed. Thus, while still predatory, they produce the smallest possible negative effect. (Thalenhorst (1958) and Lindquist (1969a,b; personal communication) suggest that the predation may be beneficial in reducing intraspecific larval beetle competition.)

A few phoretic relationships do not appear to fit the pattern so well (see, for instance, Moser, Cross and Roton 1971), and while a thorough analysis of the literature would be rewarding, it will not be attempted here. However, one system has been studied with particular thoroughness, and it is worth examining in detail.

Bark-Beetle Communities

J. C. Moser and co-workers (Moser and Roton 1971; Moser, Thatcher, and Pickard 1971; Moser 1975, 1976) have been investigating the community surrounding the southern pine bark beetle (*Dendroctonus frontalis* Zimmermann) with the goal of finding a biological control agent. The community consists of over 90 species of mites and 90 species of insects, although these numbers can be reduced somewhat by excluding the rare and the transient. Every two weeks in summer and every four weeks in winter, a beetle-infested tree was felled and 18-inch-long bolts cut at 7-foot intervals. The bark of one-third of every bolt was inspected for insects and mites within 24 hours after cutting. The remainder was placed in outdoor rearing cages, and all emerging insects collected. Moser and Roton (1971) give a full account of methods and data recorded. For our purposes, two things are of interest: (1) All observations of phoresy were recorded, both on *D. frontalis* and its other insect associates, and (2) of 90 bolts examined, *D. frontalis* was the only bark beetle in 47, while the remaining 43 contained *D. frontalis* plus other species of scolytid bark beetles. The presence of every mite species in these two classes was expressed as a ratio: (*D. frontalis* only)/(*D. frontalis* plus other bark beetles). A low number for this ratio would indicate that the mite associates primarily with other species of bark beetles, while a value around "1" would indicate that it associates primarily with *D. frontalis* or with both *D. frontalis* and other scolytids. In other words, the ratio can be used as a crude index of association.

In a separate study, Moser (1976) captured *D. frontalis* in sticky traps as they were flying toward infested trees, and recorded the frequency and species composition of phoretic mites. This study augments Moser and Roton's 1971 paper and provides valuable information on the relative abundance of different mite species on the beetle.

Finally, in a third study Moser (1975) tested 51 species of mites as predators against *D. frontalis* in the laboratory. Female, male, and pre-adult mites were separately enclosed in small chambers with eggs, first instar larvae, last instar larvae and pupae of *D. frontalis*. Predatory activity during a three-day period was then recorded. It should be stressed that predation under these circumstances does not necessarily imply predation in the field. For instance, Moser rarely sees mites in the pupal chambers, and one species (*Macrocheles boudreauxi* Krantz) that readily fed upon *D. frontalis* in the tests has since been found to feed on alternative prey if given the choice (Kinn and Witcosky 1978; Moser, personal communication). Nevertheless, the results of the tests may be taken as a crude index of predatory activity.

Now we are in a position to relate predatory activity, phoretic habits, and the general degree of association. Table 5.2 synthesizes the results of the three papers. Once again, the association index gives the number of bolts in which the mite was found with *D. frontalis* only, divided by the number of bolts in which the mite species was found with *D. frontalis* plus other scolytids. The phoresy column gives the proportion of all phoretic mites represented by that species (t = "trace," < 0.01). The predation columns give the effect of females, males, and subadult mites on eggs, first instar larvae, last instar larvae and pupal *D. frontalis*. A "0" indicates no feeding activity. A "1" indicates that the prey was less than one-fourth eaten. A "2" indicates that the prey was aggressively attacked and substantially eaten. An average score for each species is provided in the last column. Of the 51 mite species tested, 32 showed predatory activity. Species for which no index of association was available are omitted from Table 5.2.

TABLE 5.2 **Index of association (Assoc.), Phoretic tendency (Phor.), and predatory activity for 35 species of mites. Female, male, and subadult mites were tested separately with eggs (E), first instar (F), last instar (L), and pupal (P) *D frontalis*. A "0" indicates no predatory activity. A "1" indicates partial feeding. A "2" indicates aggressive predation and substantial feeding. The last column (AV) gives the average predatory score for the species. A "t" in the phoresy column indicates that the mite was only rarely transported by the beetle. (Data from Moser 1975, 1976, Moser and Roton 1971)**

Mite Species		*Phor.*	*Female* E	F	L	P	*Male* E	F	L	P	*Subadult* E	F	L	P	*AV.*
Ameroseius longitrichus Hirschmann	0/4		0	0	0	0	0	0	0	0	0	0	0	0	0
Lasioseius denatatus Fox	1/10		2	—	—	—	2	—	—	—	—	—	—	—	2.0
Proctogastrolaelaps libris McGraw and Farrier	2/3	t	—	—	—	0	—	—	—	—	—	—	—	—	0
P. dendroctoni Lindquist and Hunter	29/22		0	2	0	1	0	2	1	1	0	0	0	0	.58
P. fiseri Samsinak	0/5		0	2	2	0	0	2	2	0	0	2	2	0	1.0
P. hystricoides Lindquist and Hunter	6/13	t	0	1	0	0	0	1	0	0	0	0	0	0	.17
Dendrolaelaps isodentatus Hurlbutt	4/15		2	2	0	0	2	2	0	0	2	2	0	0	1.0
D. neocornutus Hurlbutt	6/9	t	2	2	0	0	2	2	0	0	2	2	0	0	1.0
D. neodisetus Hurlbutt	35/33	t	0	1	0	1	0	1	0	0	0	1	0	0	.33
D. rotoni Hurlbutt	1/1		2	2	1	0	2	2	1	0	2	1	1	0	1.25
D. varipunctatus Hurlbutt	2/2		0	2	2	0	0	2	2	0	0	2	0	0	.83
Longoseius cuniculus Chant	2/1	t	0	0	0	0	—	—	—	—	0	0	0	0	0
Androlaelaps casalis Berlese	0/1		2	—	—	—	2	—	—	—	—	—	—	—	2.0
Hypoaspis disjuncta Hunter and Yeh	1/0		—	—	—	—	2	—	—	—	—	—	—	—	2.0
Pseudoparasitus thatcheri Hunter and Moser	2/5		1	1	0	0	—	—	—	—	—	—	—	—	.5
Macrocheles boudreauxi Krantz	14/20	t	2	2	0	2	2	2	0	2	2	2	1	0	1.42
Eugamasus lyriformis McGraw and Farrier	9/18	t	0	2	2	2	0	2	2	2	0	2	2	0	1.33
Gamasolaelaps subcorticalis McGraw and Farrier	2/4		0	2	2	0	0	1	1	0	0	0	0	0	.5
Cercoleipus coelonutus Kinn	1/5		0	2	0	0	0	2	0	0	0	2	0	0	.5
Pleuronectocelaeno drymoecetes Kinn	2/9		2	2	1	0	1	2	1	0	1	0	0	0	.83
Histiogaster arborsignis Woodring	26/34	t	2	2	2	2	1	1	0	0	2	2	2	2	1.50
H. rotundus Woodring	6/6		2	2	2	2	2	2	2	2	2	2	2	2	2.0
Tyrophagus putrescentiae Schrank	18/12		0	0	0	0	0	0	0	0	0	0	0	0	0

(continued)

TABLE 5.2 (Continued)

Mite Species		Phor.	Female E	F	L	P	Male E	F	L	P	Subadult E	F	L	P	AV.
Anoetus insolita Woodring and Moser	3/1		0	0	0	0	0	0	0	0	0	0	0	0	0
A. sordida Woodring and Moser	4/9		0	0	0	0	0	0	0	0	0	0	0	0	0
A. varia Woodring and Moser	1/5	t	0	0	0	0	0	0	0	0	0	0	0	0	0
Pyemotes parviscolyti Cross and Moser	<.1		2	2	2	2	0	0	0	0	0	0	0	0	.67
Pygmephorellus bennetti Cross and Moser	2/17	t	0	0	0	0	0	0	0	0	0	0	0	0	0
Heterotarsonemus lindquisti Smiley	13/12	t	0	0	0	0	0	0	0	0	—	—	—	—	0
Iponemus calligraphi calligraphi Lindquist	0/1		2	0	0	0	0	0	0	0	0	0	0	0	.17
I. confusus oriens Lindquist	0/5		2	0	0	0	0	0	0	0	0	0	0	0	.17
Tarsonemus krantzi Smiley and Moser	24/19	.45	0	0	0	0	0	0	0	0	0	0	0	0	0
T. ips Lindquist	33/28	t	0	0	0	0	0	0	0	0	0	0	0	0	0
Ereynetoides scutulis Hunter	44/41	t	0	0	0	0	0	0	0	0	0	0	0	0	0
Trichouropoda australis Hirschmann	22/25	.54	0	0	0	0	0	0	0	0	0	0	0	0	0

The first prediction relates to phoresy. The theory of structured demes predicts that those mites that are dependent upon *D. frontalis* for transport will not prey upon it. Moser and Roton (1971) and Moser (1976) found 15 species of mites phoretic upon *D. frontalis*. However, two species formed the vast majority, *Tarsonemus krantzi* Smiley and Moser, and *Trichouropoda australis* Hirschmann. Neither prey upon their carrier. The method of catching beetles may have underestimated the prevalence of *Anoetus varia* Woodring and Moser, *Tarsonemus ips* Lindquist, *Dendrolaelaps neodisetus* Hurlbutt, and *Macrocheles boudreauxi* Krantz (Moser 1976). The first two are not predatory on *D. frontalis*. The third is lightly predatory (average score = 0.33). The fourth is an aggressive predator in the tests, but has since been found to prefer other prey when they are available (Kinn and Witcosky 1978). In short, although more than half the mite species that were tested prey upon *D. frontalis*, the vast majority of phoretic mites do not.

The second prediction relates to the index of association. By forming galleries under the bark, *D. frontalis* creates a habitat for dozens of other species of mites and insects. The species that are confined to this habitat depend upon *D. frontalis* just as much as its phoretic passengers do. Hence, the theory of structured demes predicts that the closest associates of *D. frontalis* will evolve positive relationships with it and, in particular, will not prey upon it. The observed relationship between the average predation index and the index of association is presented in Figure 5.1. Only species with sample sizes ⩾ 15 for the index of association are included. The results are as expected; mite species with a low index of association possess the full range of predatory activity on *D. frontalis*. However, all close associates show low predatory activity. Not a

FIGURE 5.1 Predatory ability in relation to closeness of association. Mites that associate primarily with other species of bark beetles show a full range of predatory ability on *D. frontalis*. However, mites that associate primarily with *D. frontalis* do not prey upon it. Only mite species with a sample size of ⩾ 15 for the index of association are included. (Data taken from Table 5.2.)

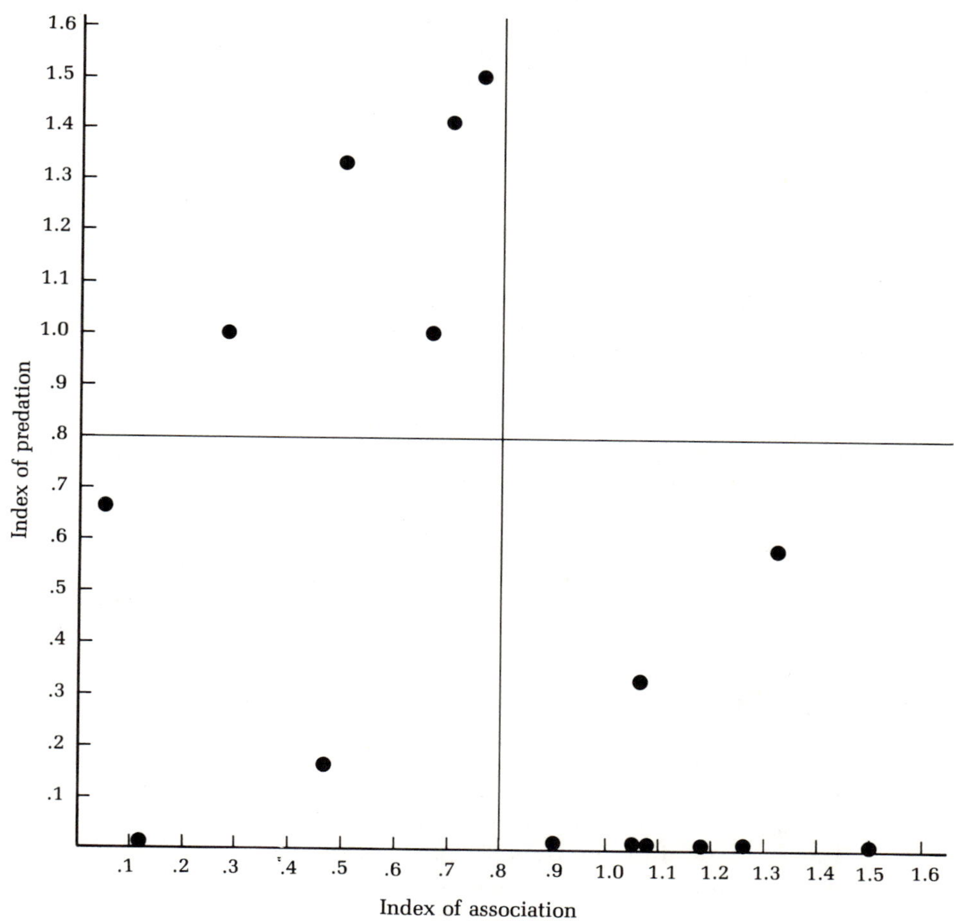

single data point exists in the upper right-hand quadrant of Figure 5.1. Mites with an index of association > 0.8 feed significantly less on *D. frontalis* than the others (t-test, d.f. = 14, t = 4.63, $p < 0.005$). When all the species in Table 5.2 are included, the difference is only marginally significant (d.f. = 33, t = 1.56, $.1 < p < 0.05$). However, mites with an index of association > 1.0 still feed much less on *D. frontalis* than the others (d.f. = 33, t = 3.01, $p < 0.005$). It should be emphasized that this pattern is not a simple artifact of destructive predators eliminating *D. frontalis* from their trait group. *D. frontalis* was abundant in all bolts examined. The low index of association is determined not by the absence of *D. frontalis*, but by the presence of other species of bark beetles.

This data is the second indication that species evolve to be less destructive on other members of their community that they depend upon (the first was the tiger-beetle study discussed in Chapter 3), and the results are encouraging because they come from a community characterized by large trait groups and many species. Aside from the theory of structured demes, I can think of only one other hypothesis to explain it. Pimentel (1961), Rosenzweig (1973), and Slobodkin (1974) have argued that if a predator and prey are locked into a coevolutionary race, prudence can still result from the prey evolving faster than the predator. But surely this is implausible, for the low associates—species that aren't even in the coevolutionary race—are superior as predators to those that are! On this point, the two hypotheses make opposite predictions. The structured-deme model predicts that the closest associates will be the least virulent, causing a paucity of data points in the upper right-hand quadrant of Figure 5.1. The coevolutionary race hypothesis predicts that closest associates should be most adept at overcoming the defenses of their prey, leading to a paucity of data points in the upper left-hand quadrant. This latter prediction is not observed. If the most closely coevolved species of mites do not prey upon *D. frontalis*, they must benefit from the association in another way, e.g., the indirect benefit of using the beetles' habitat or the beetles themselves for transport. I predict that if the biology of these mites is investigated further, they will be found to enhance the fitness of *D. frontalis*.

To summarize, an organism such as a mite has a wide array of possible resources. Through evolutionary time, it may become a feeder on plants, fungi, nematodes, insect eggs, larvae or adults, other mites, microorganisms, or these resources in combination (Lindquist 1975). At any given time, all of these ecological types, in the form of different species, are probably drifting through the habitat and will become established if the conditions are favorable enough. Traditional theory asks which of the options are most profitable, in terms of calories obtained and direct risks taken, and stops there.

But if indirect effects are taken into account, this kind of analysis clearly becomes inadequate. In my opinion, the patterns of predation upon the bark beetle cannot be understood without recognizing that the bark beetle performs useful functions for its community by building galleries under the bark and serving as an agent of transport. Similarly, an understanding of the mites requires a knowledge of function—for the same evolutionary force causing them to decrease their negative trophic effect on the bark beetle will cause them to acquire positive nontrophic effects, as we have seen in carrion-beetle communities (Springett 1968).

Traditional theory does not sufficiently emphasize the extent to which the environment is created by the activities of organisms, and how completely the fitness of an individual can depend on other members of the community. A bark beetle does not simply eat trees. It creates a whole little world. Its activities provide a favorable microclimate, protection from many predators, and access to a resource. The beetle modifies the structure of the environment, which affects the fitness of many other species.

These other species also do not simply eat things. They too have activities that modify the habitat and alter the rates at which other processes occur. Consider, for instance, the possible activities of yet another segment of bark-beetle communities—the fungi growing within the galleries. Fungi can differ in growth form, either blocking passages or hugging the walls of the galleries. They can alter the microclimate, especially the moisture content and the tendency of sap to flood the chambers. They can alter the chemical environment, by breaking down tannins or producing toxins of their own. They can convert unavailable resources to a form that is usable by bark beetles. They can facilitate or inhibit the growth of other fungi with any of the preceding properties.

What kinds of fungi are likely to be found in bark-beetle communities? It is impossible to answer this question without considering the activities of other members of the community, and how they relate to the fungi. Of course, because the bark beetle creates the habitat in the first place, there should be an evolutionary initiative on the part of the fungi to enhance the fitness of the beetle, exactly as for the mites. But in addition, there are other activities by which the community can alter fungal fitness. Dispersal between habitats is carried out largely on the bodies of animals, and can occur differentially. Obviously, animals can cause differential mortality within the galleries. They can even cause the differential mortality of unpalatable species by a process analogous to the human activity of weeding. Animals can also secrete substances that cause the differential growth of fungi.

Given the evolution of indirect effects, coupled with the power of the bark-beetle community over its members, it would not be surprising if a positive function became the ultimate criterion of selection—the way to maximize individual fitness. The major species of animals and fungi do indeed seem to form positive relationships in bark-beetle communities, utilizing all of the pathways discussed above (reviewed by Graham 1967; see also Batra 1966, 1972, and Batra and Batra 1967).

Are bark-beetle communities unusual in their tendency to evolve into superorganisms? Are there any special conditions that are not likely to be met in other systems? The two essential ingredients are the evolution of indirect effects and a large degree of dependency between species. Bark-beetle communities are colonized by many individuals of many species. The animals are attracted to the tree by pheromones released by the first bark beetles to arrive (Borden 1974). As such they are recruited from a wide area, and although some colonists of the new habitat may be derived from the same parent gallery, there is no reason to expect a high genetic similarity between colonists in general. These are stringent conditions for the structured-deme model, yet the evolution of indirect effects is still observed. Many other systems in nature must possess an equivalent or a more favorable degree of trait-group variation.

The dependency between species and the degree to which dependency can be molded by natural selection are open questions that must be decided by future study. Some species (such as the ripple bug discussed in Chapter 3) might be unaffected by the rest of the community. However, many others must be as dependent upon their neighbors as are the inhabitants of bark-beetle galleries. The suitability of soil as a growth medium for plants must surely be determined by the activities of organisms, and coral reef ecosystems—among the most fantastic in the world—are constructed in the midst of both a structural and a nutritional desert (Muscatine and Porter 1977). Furthermore, it is possible that dependency between species evolves as an active process. In fact, structured-deme theory makes just such a prediction.

The Concept of Dependent Specialization

Most ecological discussions of specialization consider a continuum of resources and a set of species utilizing various segments of the continuum. Specialization is related to the narrowness of the segment that is utilized. Thus, a bird that forages only on the trunks of trees is more specialized than a bird that gleans foliage and catches insects on the wing, in addition to trunk foraging. A herbivore that feeds on one plant species is more specialized than a herbivore feeding on many species (see Pianka 1974 for a representative treatment).

Specialization presumably exists because a small number of activities can be performed more efficiently than a large number of activities. Thus, a bird that forages only on trunks can devote its entire genome to that one task. Its feet evolve to be especially adept at clinging to vertical surfaces, its beak adept at tearing off bark, and its tongue adept at probing crevices for insects. These adaptations actually decrease fitness in other environments, but on tree trunks the trunk specialist has no equal. If every resource has such a specialist, a generalist will find it impossible to compete.

I would like to distinguish between two types of specialization: (1) the specialization of *tasks*, which refers to the basic process whereby one activity can be performed more efficiently than many, and (2) the specialization of *needs*, which refers to the variety of substances, or resources, required by an organism to live.

The relation between these two types of specialization can best be described in the context of human communities. For humans, the specialization of tasks proceeds to an almost absurd degree. Apparently the simplest activities can be done more efficiently if they are made even simpler—witness assembly lines. As another example, pins in the eighteenth century were most efficiently manufactured by eleven distinct operations, each performed by a different person (Smith 1776).

Yet the potential for specialization of needs is very limited. Everyone needs a variety of foods, clothing, shelter, and a number of tools. It is clear that if everyone had to fulfill all their own needs, human societies would never advance beyond the subsistence farming stage. Yet, specializing further requires becoming *less* proficient at providing some of the necessities of life!

The problem is obviously solved by a community of complementary specialists in which each person concentrates on a small subset of necessities, and the products of each person's labor are made available to the rest of the community. Only in this way can the increased efficiency inherent in the specialization of tasks be realized. In short, if the specialization of tasks is to proceed beyond the specialization of needs, an increase in dependency of individuals on their community is inevitable. This concept may be termed "dependent specialization."

What is required for the evolution of dependent specialization in nature? Consider a community composed of three species (S_1, S_2, S_3), each of which requires two resources (taken in the broadest possible sense). In other words, all three species have identical needs. Whichever resource is in the shorter supply limits the growth of the populations. Furthermore, the species not only require the two resources, but are responsible for their production through various activities or modifications of the environment. Species 3 is a generalist that attempts to promote both resources at once. Species 1 and 2 are specialists on resources 1 and 2 respectively. Because of the specialization of tasks, they are more efficient than the generalist at cultivating the resources.

At equilibrium, the abundance of the two resources may be expressed as

$$R_1 = U_1p_1 + U_3p_3 \quad \textbf{(5.2)}$$

$$R_2 = U_2p_2 + U_3p_3 \quad \textbf{(5.3)}$$

where R_i is the abundance of resource i, p_i is the proportion of species i in the community, and U_i is the resource abundance that would exist if the population were composed entirely of species i.

Whichever resource is in shorter supply determines the carrying capacity (K) for the community

$$K = hR_1 \quad \textit{if} \quad R_1 < R_2 \quad \textbf{(5.4)}$$

$$K = hR_2 \quad \textit{if} \quad R_2 < R_1 \quad \textbf{(5.5)}$$

where h is a constant. Notice that for the specialists to exceed the efficiency of the generalist, U_1 and U_2 must each be greater than $2U_3$. To see this, let $U_1 = U_2 = 100$, $U_3 = 50$, and $h = 1$. A community composed entirely of species 3 (the generalist) would have a carrying capacity of 50. A community composed of 0.5 species 1 and 0.5 species 2 (the specialists) would also have a carrying capacity of 50. Any other proportionality of the specialists decreases the carrying capacity, which is determined by the resource in shortest supply. Hence, for specialization to be adaptive, the increased efficiency on one task must overcompensate for the decreased efficiency on the other.

Because the species do not differ in their utilization of the resources, their proportionalities within a single trait group cannot change, and at equilibrium,

$$N_i = p_iK \quad \textbf{(5.6)}$$

where N_i is the density of species i. Individual selection models, therefore, do not predict the evolution of dependent specialization.

This is a fairly fundamental conclusion. Since dependent specialization necessarily involves a reciprocal transfer of products of specialization between members of the trait group, it is difficult to see how it can increase the relative fitness of the specialists within the trait group without sophisticated behavioral structuring unlikely to be found in most organisms (Trivers 1971). However, in structured demes the trait groups vary in their proportions of specialist vs. generalist. Those trait groups with a high proportion of both specialists are most successful at cultivating the resource. They produce more individuals than trait groups inhabited by generalists and, therefore, make a differential contribution to the deme.

A single example of this process is shown in Figure 5.2 (see Chapter 4 for details of the computer simulation). Here, $U_1 = U_2 = 150$, and $U_3 = 50$. Thus, a trait group composed entirely of generalists has a carrying capacity of 50, while a trait group composed of equal proportions of specialists has a carrying capacity of 75. The specialists are initially rare relative to the generalist. The variance in density between trait groups is random, i.e., equal to the mean; and the covariances between densities are 0. The specialists replace the generalist in approximately 400 generations. While this is a long period of time, it must be remembered that the parameter values used in this example are very stringent for the structured-deme model. The trait groups are large (50 to 75), the trait-group variation is random, the covariances are 0, and the advantage to specialization is relatively slight. If $U_1 = U_2 = 30$ and $U_3 = 10$ (equals a trait-group size of 10 to 15), an equivalent amount of selection occurs in roughly 50 generations.

The structured-deme model thus predicts that in biological communities the specialization of tasks can routinely evolve beyond the specialization of needs, resulting in an increased interdependency between species. This is a *functional* interdependency. Each species performs an activity that benefits the community, and in return is supported by activities the species is incapable of performing. In a sense, a species' relation to the community resembles an organ's relation to the organism—a part of a larger whole, designed to serve the larger whole, not an adaptive entity by itself.

This model does not take into account situations where specialists find themselves without complementary specialists (although the stochastic process that creates variation among trait groups does generate imbalances in some of the trait groups, which in a sense are necessary to drive the process). Such situations will definitely inhibit the evolution of dependent specialization. So also will situations in which it is difficult to transfer the products of

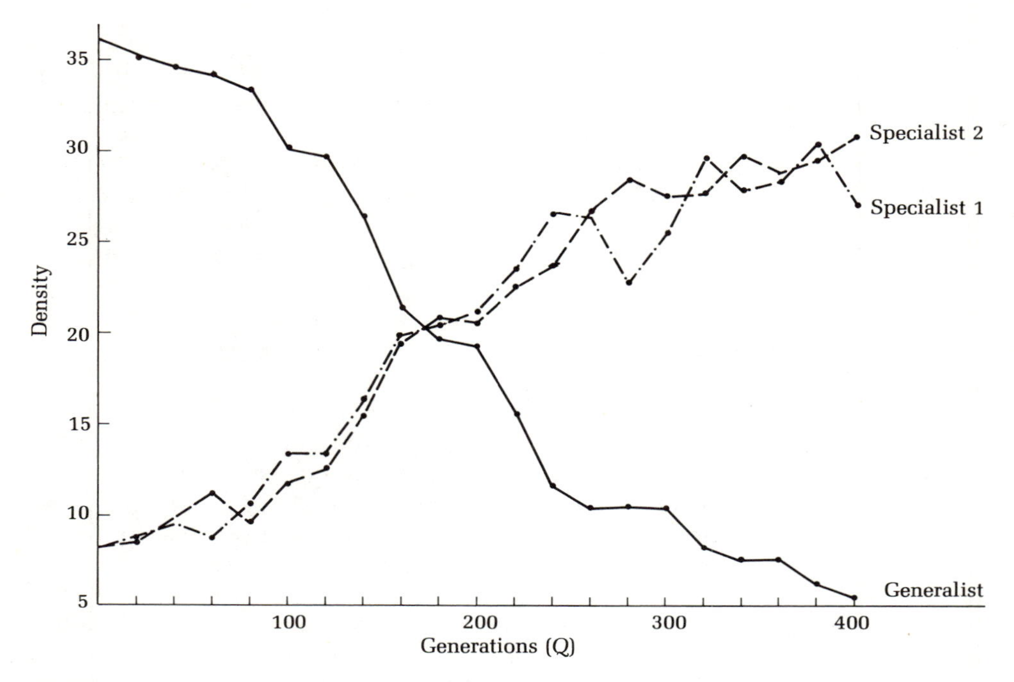

FIGURE 5.2 A single simulation run of dependent specialization model in structured demes. Each specialist is incapable of existing by itself, but both together outcompete the generalist.

specialization. I do not wish to assert that biological communities are tightly organized superorganisms—that is an open question. I suggest only that the concept of a species performing a function for its community can be reconciled with evolutionary theory.

A Possible Method of Biological Control

The models presented in this book need to be refined and tested before they can be trusted. Nevertheless, it is interesting to speculate on a possible application to biological control. Consider a diverse community of plants and animals, one species of which is a pest, i.e., undesirable from the human standpoint. Other species in the community may be divided into three categories with respect to their effect on the pest:

1. Those species that are beneficial to the pest, enhancing its presence through any pathway (e.g., as a resource, a refuge, a microclimate modifier, or by enhancing another species that more directly benefits the pest).

2. Those species that have no effect on the distribution and abundance of the pest.
3. Those species that are antagonistic to the pest, inhibiting its presence through any pathway (e.g., as a predator, a competitor without itself being a pest, a destroyer of the microhabitat, or an antagonist to a beneficial species).

Now consider the small-scale spatial distribution of the pest. If it is like most species, it will have a patchy distribution with local areas of rarity and abundance. This patchiness will have two components:

1. A stochastic component; the pest may simply have colonized some areas but not others, it may have been concentrated by the wind, or an unpredictable disturbance may have eliminated the pest in one area but not another. For reasons that will become apparent, this may be termed the "nonheritable" component of spatial distribution.
2. A biotic component; the distribution of the pest may be influenced by the underlying distribution of beneficial and antagonistic species in the community. In other words, the pest will be abundant where its benefactors are abundant and rare where its antagonists are abundant. This may be termed the "heritable" component of spatial distribution.

We desire to eliminate the pest. One approach is to apply a poison over the entire landscape that is toxic only to the pest. This approach will accomplish a short-term reduction, but the community remains a favorable place for the pest to regrow. Alternatively, consider the consequences of employing a very general but short-lived toxin *only on areas of local pest abundance*. The toxin would eliminate a large segment of the total community (for instance, all arthropods), but would be harmless to individuals colonizing the area from adjacent patches, a few meters away, a few days later. This latter method also accomplishes a short-term reduction in pest numbers; but by applying community-wide mortality in a way that correlates with pest spatial distribution, we are also changing the fundamental structure of the community with respect to the pest. If there is a heritable component to the pest spatial distribution, then the toxin will be differentially applied to all members of the community that benefit the pest. Patches made vacant by application of the toxin will be recolonized by organisms from adjacent patches. Since the pest was not common in the adjacent patches, the species composition of the colonists is biased toward the antagonistic component of the community. By repeating the procedure, it seems possible to cause a shift in species composition, creating a more hostile community for the pest. Over the long term, the pest should decrease in abundance and the toxin could be applied less often. In a sense, we have selected for a community that eliminates the pest for us. Notice that a detailed knowledge of each species' biology is not necessary. Sufficient information is contained in the spatial distributions of the pest and in its necessary correlation with the distribution of benefactors and antagonists.

This method of biological control requires a large species diversity and an assessment of small-scale spatial distribution that makes it unsuitable for many agricultural systems, yet it may still hold promise for some crops. For instance, mushroom farming involves making compost as sterile as possible, spreading it onto trays, seeding it with mushroom mycelia, harvesting after a period of time, and repeating the procedure. During mushroom growth the trays are colonized by a diverse array of microbes, other species of fungi, nematodes, mites, and insects, some of which are serious mushroom pests. It would be fascinating to introduce one alteration to the procedure: Take note of the spatial distribution of mushrooms on the tray. Perhaps areas of prolific growth are caused by the presence of beneficial species, that directly facilitate the mushroom or keep out mushroom pests. Take a small amount of compost from the most productive patches and mix it in with sterilized compost for the next crop. This addition will give the beneficial organisms a head start as colonists. With many iterations it may be possible to develop a community that is, in a sense, built and designed to facilitate mushroom production.

As another example, Room (1971, 1973, 1975) has investigated the relationship between pest abundance and ant-species distributions on tropical tree crops, such as cacao. The ant species differ in their effect on the pest, yet still exclude each other from their foraging territories. By eliminating the ant fauna from trees infested with pests, those trees would be available for occupancy by colonies from adjacent trees that have a more negative effect on the pests. Since "possession is nine points of the law" for many forms of interspecific competition, the altered distributions could be stable, i.e., the undesirable ant species could be permanently excluded from the area.

Although the foregoing discussion is obviously speculative, at least it demonstrates the utility of viewing the multispecies community as an adaptive unit. The proposed method of biological control involves measuring a phenotypic trait (pest-species abundance) for a large number of communities. We then select the communities at the lowest end of the phenotypic range. These are the "parents" from which the next generation of communities is derived. If the phenotypic variation has a heritable component, then its mean will shift in the direction of selection over time. The average community will have a lower pest density, caused by antagonistic relationships with other members of the community. The community will take on characteristics that are especially adapted to exclude the pest. What better term exists for this process than "the artificial selection of communities"?

The artificial selection of communities involves applying an external source of mortality in a way that spatially correlates with pest abundance. In a sense we make it group-advantageous for each species to exclude the pest from its trait group, for by doing so it escapes trait-group extinction. Now let us return to untampered biological communities and consider a species that itself has a negative effect on other members of the community. Nothing has changed. The mortality simply emanates from the species itself instead of from an external source. What better term exists for this process than "the natural selection of communities"?

6 Multilevel Evolution

Individual Welfare and Community Welfare

The problem of group selection and community evolution fundamentally begins with a conflict between the welfare of the whole system and the welfare of some of its elements. This conflict is real. There is abundant evidence for it in human communities, single-species populations, and multispecies communities; a large mathematical literature is devoted to its exploration (e.g., Shubik 1975). Of course, systems can be constructed in which every member contributes to the welfare of the whole, but they require a very special kind of structure that cannot be automatically expected in nature. For our purposes, it is safe to assume that most arbitrary collections of individuals will not interact in a way that maximizes community fitness. For the evolution of the superorganism, some organizing force must be present, suppressing the activity of certain individuals and enhancing that of others.

Theoretically, there are two ways that such organization can occur. The first is for every individual to be directly sensitive to community welfare, voluntarily curtailing its own activity whenever detrimental to the whole system. While this would indeed resolve the conflict between individual welfare and community welfare (by simply ignoring the former), it is impossibly at odds with the mechanism of evolution, so strongly based on advantages to individuals. Theories of altruism can explain some examples of voluntary curtailment, but even the staunchest believer of altruistic behavior would not propose that individual self-interest is completely abandoned. One concludes that direct sensitivity to community welfare is a weak organizing force, occurring only to the extent that altruism occurs.

In this book I have attempted to construct an alternative paradigm, more consistent with principles of self-interest. Every member of the community is assumed to be sensitive only to its own welfare—evolving positive responses to positive effects from other members of the community and negative responses to negative effects. While this is purely selfish behavior in the intuitive sense of the word, it involves the operation of indirect effects and, therefore, the mechanism of structured demes, or an equivalent amount of behavioral structuring in the form of individual recognition.

Indirect effects are thus postulated to be the unit interaction of community evolution. Over evolutionary time, the effect of the community on a member comes to reflect the effect of that member on the community, and only those members that positively contribute to community welfare persist. This is the actual process of community evolution. It was stated as an analogy to human communities in Chapter 4 and is now explored more carefully here.

Before beginning, however, it is important to note a corollary. If selfish genes organize themselves into selfish individuals, selfish individuals into selfish populations, and selfish populations into selfish communities, then evolution operates on all levels. It makes no more sense to say that the gene (Dawkins 1976) or the individual (Mayr 1963) is the fundamental unit of selection than to say that the molecule or the cell is the fundamental unit of physiology. All this rests on a demonstration that communities whose members are motivated entirely by self-interest evolve into superorganisms.

Simple Linear Model

As a prelude to a more realistic model, consider a community composed of a number (S) of species, whose densities ($N_1, N_2, \ldots, N_s$) are governed by the following equations where a_{ij} is the per capita effect of species j on species i:

$$\begin{bmatrix} N_{1,t+1} \\ N_{2,t+1} \\ \cdot \\ \cdot \\ \cdot \\ N_{s,t+1} \end{bmatrix} = \begin{bmatrix} 1 & a_{12} & \ldots & a_{1s} \\ a_{21} & 1 & \ldots & a_{2s} \\ \cdot & & & \\ \cdot & & & \\ \cdot & & & \\ a_{s1} & a_{s2} & \ldots & 1 \end{bmatrix} \begin{bmatrix} N_{1,t} \\ N_{2,t} \\ \cdot \\ \cdot \\ \cdot \\ N_{s,t} \end{bmatrix} \tag{6.1}$$

Needless to say, as a model of biological communities, this one is hopelessly unrealistic. There are no carrying capacities, so given a sufficient length of time, the densities explode to plus or minus infinity. Nothing is specifically labeled a predator, competitor, or prey (or anything else), and the nonlinearities that characterize any realistic biological model are absent. However, this model does capture the one essential quality of biological communities that we want to consider, in the simplest possible form: Every species depends for its existence on interactions with the other species in the community.

The evolution of this community can be explored with a computer simulation, using the following procedure:

1. Starting densities ($N_{i,o}$) are specified, and random numbers (a_{ij}, $j \neq i$) uniformly distributed between $+u$ *and* $-u$ are entered into the matrix ($u = 0.1$ in the following simulations).
2. The equations are iterated for as long as desired. To keep numbers within reasonable bounds, the sum of the species densities (hereafter referred to as the community density) is normalized to a value of 50 after every iteration. The species densities before normalization follow some trajectory through time. Species with negative densities are set to 0. The most interesting parameter is the community density because it indicates whether the totality of interactions has been positive or negative. This step serves as a control run, giving the behavior of the system for a specified set of matrix-element values.
3. An experimental run, in which the matrix elements "evolve," is initiated. It has exactly the same starting value and procedure as the control run, except that after every iteration the matrix elements are changed according to the rules of indirect effects

$$a_{ij,t+1} = a_{ij,t} + e_{ij,t} \quad \textbf{(6.2)}$$

$$e_{ij,t} = z a_{ji,t} N_{i,t} \quad \textbf{(6.3)}$$

where z is a constant, chosen such that $e_{ij,t}$ is small relative to $a_{ij,t}$ ($z = 0.01$ in the simulations). As an example, if species i has a negative effect on species j, ($a_{ji}N_i < 0$), species j evolves its effect on i to be slightly more negative than it was before. Every species evolves to inhibit its enemies and stimulate its

> allies. Through this process some species are enhanced and others are inhibited, compared to the control run. However, the parameter of greatest interest is the community density because it indicates whether the net result of evolution has been positive or negative.

The results of forty simulation runs for a five-species community are shown in Figure 6.1. The solid line indicates the average difference between the control and experimental run, while the bars give the standard deviation. If the matrix happens to receive a large

FIGURE 6.1 The average difference between evolving (exptal) and nonevolving (control) communities for 20 iterations of linear model. Evolution has increased community density in 39 of 40 simulation runs. The bars indicate standard deviations.

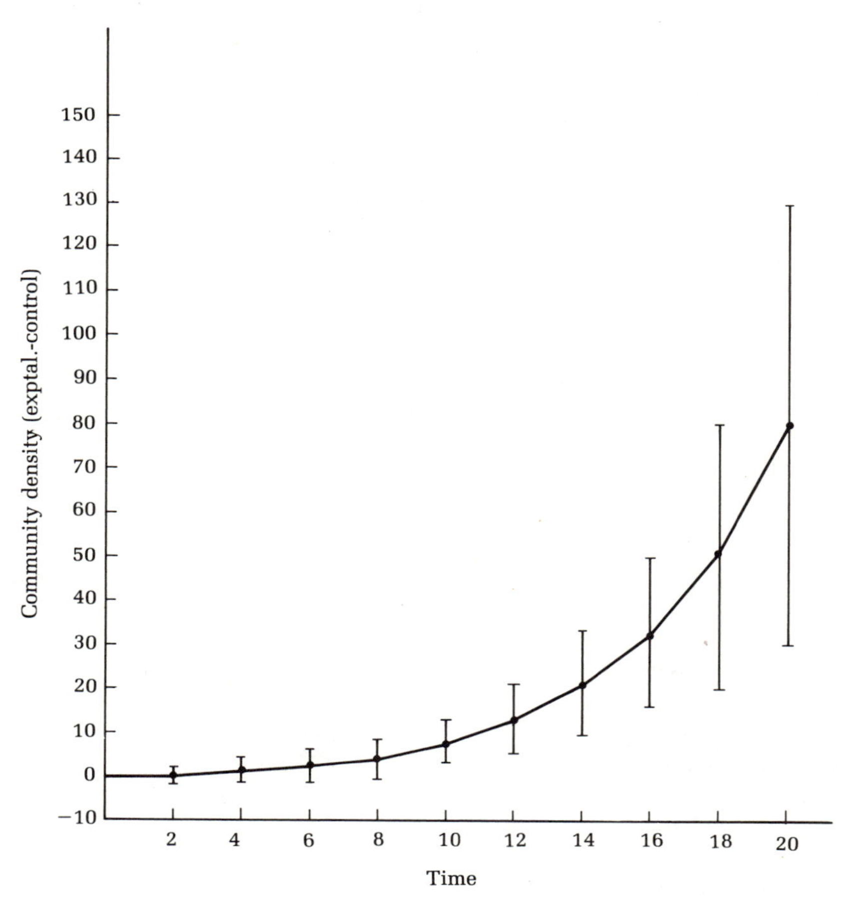

number of negative elements, the density of the evolving community initially becomes slightly lower than the control. However, in 39 of the 40 simulation runs, the density of the evolving community either immediately or eventually exceeds the control, and then proceeds to rise exponentially (it also increases in terms of absolute abundance). In the single exception, four species became extinct, leaving a single survivor that had nothing to coevolve with.

FIGURE 6.2 A single example of the linear model presented in detail. At $t = 0$, the matrix is composed of random numbers uniformly distributed between +0.1 and −0.1. The density trajectories of the species for 20 iterations is shown in the graph labeled "control." The density trajectories for the evolving community are shown in the graph labeled "exptal." After 20 iterations of evolution by indirect effects, the experimental community matrix is shown at $t = 20$.

$t = 0$

1	.036	−.071	.003	.013
.044	1	−.076	−.04	.089
−.05	.076	1	.073	−.046
.03	−.06	−.033	1	.089
.003	.066	.005	.067	1

$t = 20$

1	.162	−.197	.062	.048
.244	1	.005	−.47	1.126
−.13	.031	1	.075	−.067
.101	−.456	.131	1	1.221
.095	1.39	−.208	1.46	1

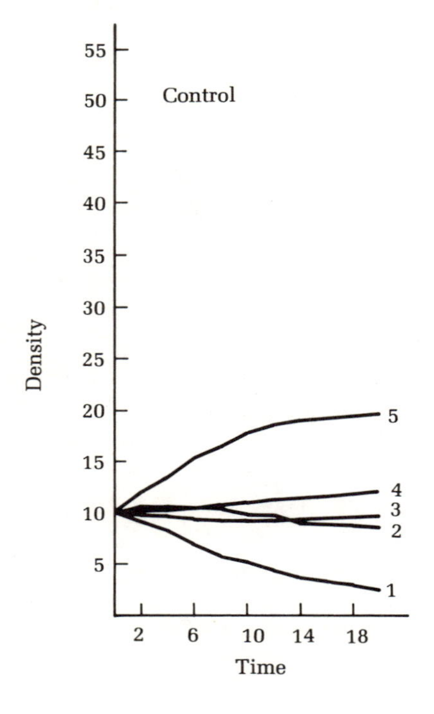

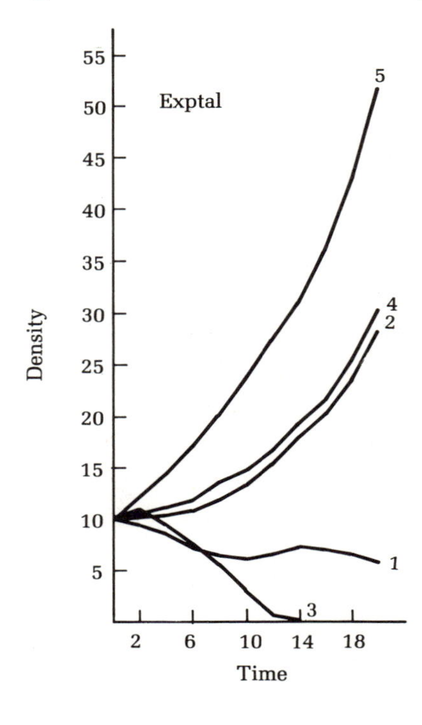

It is important to realize that these results do not follow in any simple way from the rules of indirect effects. Because the initial matrix-element values are random numbers between $+u$ and $-u$, the probability of a species having a net negative effect on its community is 0.5. Similarly, when any single species evolves its column elements by the rules of indirect effects, the results can easily be negative for the whole system (demonstrated on p. 145). Nevertheless, when evolution occurs among all species, community density increases in 39 of the 40 runs conducted.

The detailed results for a single run are provided in Figure 6.2, which allows one to trace the development of several mutualistic and inhibitory relationships. For instance, species 3 starts out with a highly negative effect on its community (column 3 of matrix at $t = 0$), which does little to inhibit its own prevalence in the nonevolving community (left-hand graph). However, in the evolving community it is quickly taken to extinction. Mutualistic associations have formed between 2 and 5, and 4 and 5; and an antagonistic interaction, between 2 and 4. The overall effect of evolution has been highly positive.

Robustness

To be certain that the principal conclusions of this model are the result of indirect effects and not artifacts of some of its many artificialities, the following modifications were made in the program and 20 simulation runs obtained from each. (More fundamental modifications that increase the biological realism of the model are presented later.)

The first modification incorporates nonlinear interactions. Most ecological models describe interactions between species as the products of their densities, as would be the case if they were moving particles passively colliding with each other. Equation (6.1) was therefore changed to

$$\begin{bmatrix} N_{1,t+1} \\ N_{2,t+1} \\ \cdot \\ \cdot \\ \cdot \\ N_{s,t+1} \end{bmatrix} = \begin{bmatrix} 1 & N_{1,t}a_{12} & N_{1,t}a_{13} & \dots & N_{1,t}a_{1s} \\ N_{2,t}a_{21} & 1 & N_{2,t,}a_{23} & \dots & N_{2,t}a_{2s} \\ \cdot & & & & \\ \cdot & & & & \\ \cdot & & & & \\ N_{s,t}a_{s1} & N_{s,t}a_{s2} & \cdot & \cdot \quad \cdot & 1 \end{bmatrix} \begin{bmatrix} N_{1,t} \\ N_{2,t} \\ \cdot \\ \cdot \\ \cdot \\ n_{s,t} \end{bmatrix} \qquad (6.4)$$

To keep the element values at roughly the same order of magnitude of the original model, u was changed from 0.1 to 0.01.

The second modification normalizes the sum of the columns. In nature it is likely that organisms have a limited potential to affect other members of their community. An increase in their effect on one species (a "specialization") must be accompanied by a decrease in their effect on others. The limited potential may be simulated by normalizing the sum of the absolute values of every column in the matrix, exclusive of the diagonals, to a given number (0.3 in the simulations), initially and again after every iteration in the evolving community. Now the evolution of indirect effects involves only a redistribution of adaptive potential, rather than a change in magnitude.

The third modification involves nonnormalized densities. To keep densities within manageable limits in the original model, the community is started at a total density of 50 and normalized back to 50 after every iteration. The community densities of Figures 6.1 and 6.2 are displacements from this standard value. The normalization procedure could conceivably affect the basic results, so it was eliminated for 20 simulations.

Figure 6.3 displays the results of the three modifications. Incorporating nonlinearities (a) produces a pattern very similar to that shown by the simple model (Figure 6.2). Normalizing the columns causes the community density to level off as the species optimally distribute their adaptive potential (b). Astronomical densities are obtained when the density normalization procedure is not used (c). However, in 59 of the 60 simulation runs, the density of the evolving community exceeds the control, exactly as in Figure 6.1 (the one exception evolved into a set of mutually negative interactions that decreased community density). Clearly, this fundamental pattern must be a result of the way the matrix elements are evolved.

The process whereby "selfish" communities evolve in the direction of increased density can be explored analytically for a simplified case. Consider a control run of the initial model in which the density normalization procedure is not used. Iterating the equations is equivalent to raising the matrix to a power, so after two iterations equation (6.1) may be rewritten:

$$\mathbf{n}_{t+2} = \mathbf{A}^2\mathbf{n}_t \tag{6.5}$$

where $\mathbf{A}$ is the matrix of interaction coefficients and $\mathbf{n}$ is the col-

FIGURE 6.3 Average difference between evolving and nonevolving communities for three modifications of the linear model. See text for explanation. →

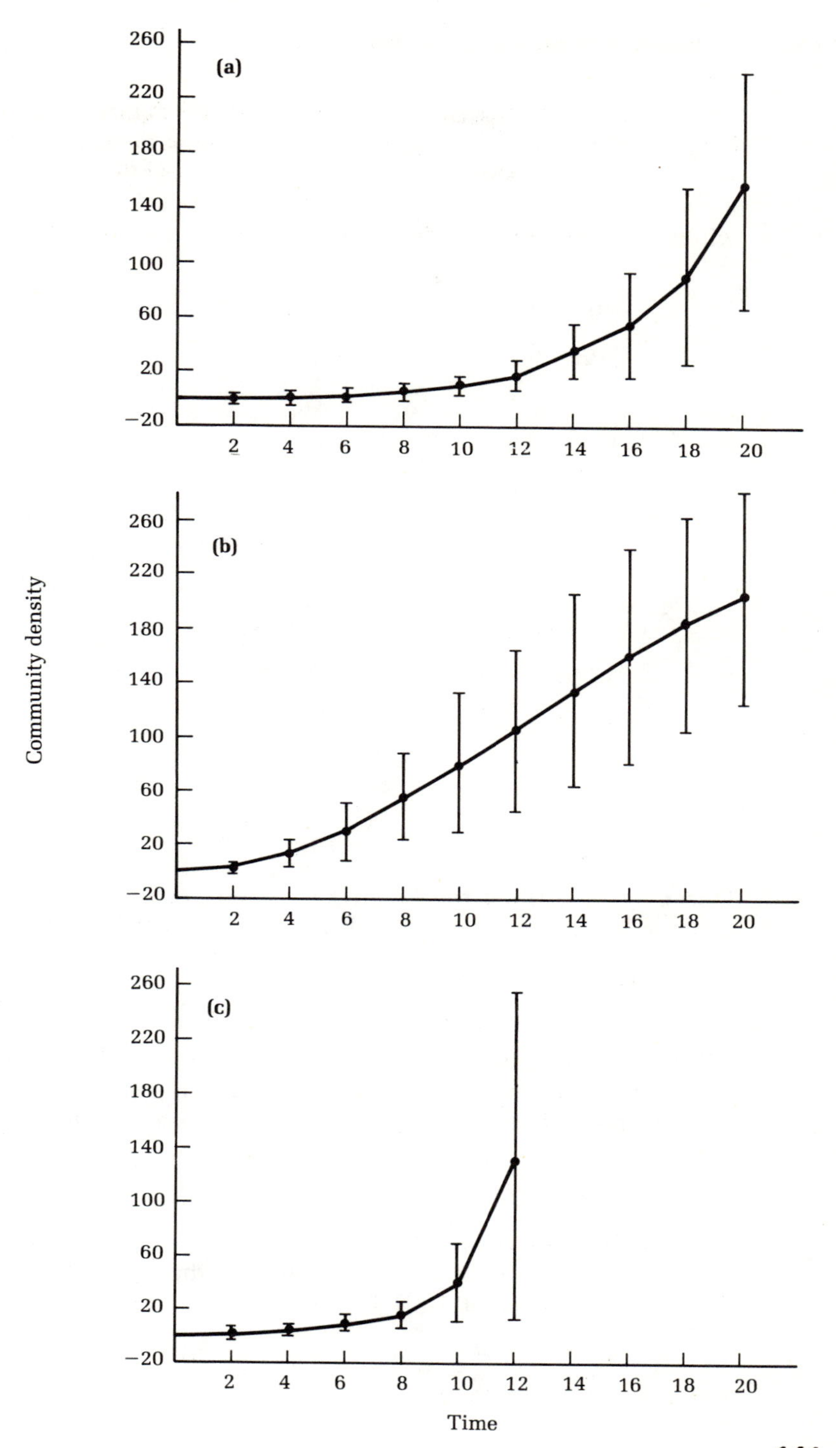
(a)
(b)
(c)
Community density
Time
260
220
180
140
100
60
20
−20
2
4
6
8
10
12
14
16
18
20

umn vector of densities. In the experimental run the elements are changed after the first iteration, yielding a new matrix

$$\hat{\mathbf{A}} = \mathbf{A} + \mathbf{E} \tag{6.6}$$

where the elements of **E** are determined by equation (6.3):

$$\mathbf{E} = \begin{bmatrix} 0 & e_{12} & e_{13} & \cdots & e_{1s} \\ e_{21} & 0 & e_{23} & \cdots & e_{2s} \\ \cdot & & & & \\ \cdot & & & & \\ \cdot & & & & \\ e_{s1} & e_{s2} & & \cdots & 0 \end{bmatrix}$$

$$= \begin{bmatrix} 0 & za_{21}N_1 & \cdots & za_{s1}N_1 \\ za_{12}N_2 & 0 & \cdots & za_{s2}N_2 \\ \cdot & & & \\ \cdot & & & \\ \cdot & & & \\ za_{1s}N_s & & \cdots & 0 \end{bmatrix} \tag{6.7}$$

This may be termed the "matrix of newly evolved effects." After two iterations, the experimental run may be written:

$$\begin{aligned} \mathbf{n}_{t+2} &= \mathbf{A}(\mathbf{A} + \mathbf{E})\mathbf{n}_t \\ &= \mathbf{A}^2\mathbf{n}_t + \mathbf{AEn}_t \end{aligned} \tag{6.8}$$

Subtracting equation (6.5) from (6.8), we see that if the sum of the elements in the column vector $\mathbf{AEn}_t$ is greater than 0, then evolution by indirect effects has caused an increase in community density

$$\mathbf{1}^T(\mathbf{AE})\mathbf{n}_t > 0 \tag{6.9}$$

where $\mathbf{1}^T$ is a row vector of ones.

Consider the specific case in which the species densities at time = t are each equal to N. Equation (6.9) then becomes

$$zN^2\mathbf{1}^T(\mathbf{A}\hat{\mathbf{E}})\mathbf{1} > 0 \tag{6.10}$$

where z and N are scalars and

$$\hat{\mathbf{E}} = \begin{bmatrix} 0 & \hat{e}_{12} & \hat{e}_{13} & \cdots & \hat{e}_{1s} \\ \hat{e}_{21} & 0 & \hat{e}_{23} & \cdots & \hat{e}_{2s} \\ \cdot & & & & \\ \cdot & & & & \\ \cdot & & & & \\ \hat{e}_{s1} & & & \cdots & 0 \end{bmatrix}$$

$$= \begin{bmatrix} 0 & a_{21} & a_{31} & \cdots & a_{s1} \\ a_{12} & & & & \\ \cdot & & & & \\ \cdot & & & & \\ \cdot & & & & \\ a_{1s} & & & \cdots & 0 \end{bmatrix} \quad (6.11)$$

$\hat{\mathbf{E}}$ is simply the transpose of $\mathbf{A}$ with zeroes in the diagonal. Equation (6.10) can be shown to equal

$$zN^2 \left[\sum_{j=1}^{s} (a_{ij} + a_{2j} + \cdots + a_{sj}) \cdot (\hat{e}_{j1} + \hat{e}_{j2} + \cdots + \hat{e}_{js}) \right] > 0 \quad \textbf{(6.12)}$$

and rearranging

$$zN^2 \left[\sum_{i=1}^{s} \sum_{j=1}^{s} \hat{e}_{ij} + \sum_{j=1}^{s} (a_{1j} + a_{2j} + \cdots + a_{sj} - a_{jj}) \cdot (\hat{e}_{j1} + \hat{e}_{j2} + \cdots + \hat{e}_{js} - \hat{e}_{jj}) \right] > 0 \quad \textbf{(6.13)}$$

$$zN^2 \left[\sum_{i=1}^{s} \sum_{\substack{j=1 \\ j \neq 1}}^{s} a_{ij} + \sum_{j=1}^{s} (a_{1j} + a_{2j} + \cdots + a_{sj} - a_{jj})^2 \right] > 0 \quad \textbf{(6.14)}$$

The second term of the bracketed expression is a sum of squares and will always be positive. The first term is the sum of the off-diagonal elements in matrix **A** (or **E**). We thus have two cases:

1. $\sum_{i=1}^{s} \sum_{\substack{j=1 \\ j \neq 1}}^{s} a_{ij} > 0.$ In this case, evolution by indirect effects always increases community density.
2. $\sum_{i=1}^{s} \sum_{\substack{j=1 \\ j \neq i}}^{s} a_{ij} < 0.$ In this case, evolution may decrease community density. We know from the simulations that community density does decrease when the matrix is highly negative—but only temporarily. After several iterations, it increases again. Unfortunately, the mathematics after one iteration become too complex to explore this reversal analytically.

The positive force that always causes an increase in community density is the sum of squares. Written in the form of equation (6.13), it may be given a biological interpretation:

$$\sum_{j=1}^{s} (a_{1j} + a_{2j} + \cdots + a_{sj} - a_{jj}) (\hat{e}_{j1} + \hat{e}_{j2} + \cdots + \hat{e}_{js} - \hat{e}_{jj}) \qquad \textbf{(6.15)}$$

Recall that $\hat{e}_{ji}$ is the newly evolved effect of species i on species j. The second factor can thus be interpreted as the newly evolved effect of the community on species j (self-effects excluded). The first factor is similarly the old effect of species j on the community. We may therefore say that *given the evolution of indirect effects, the newly evolved effect of the community on the species multiplied by the old effect of the species on the community, always increases community density.*

As a modified example, consider the situation in which only one species is allowed to evolve by indirect effects. Then

$$\hat{\mathbf{E}} = \begin{bmatrix} 0 & 0 & 0 & \ldots & 0 \\ a_{12} & 0 & 0 & \ldots & 0 \\ a_{13} & & & & \\ \cdot & & & & \\ \cdot & & & & \\ \cdot & & & & \\ a_{1s} & 0 & 0 & \ldots & 0 \end{bmatrix} \qquad \textbf{(6.16)}$$

And expression (6.14) becomes

$$zN^2 \left[\sum_{j=1}^{s} a_{1j} + \sum_{j=1}^{s} (a_{ij} + a_{2j} + \cdots + a_{sj} - a_{jj})(a_{1j}) \right] > 0 \quad \textbf{(6.17)}$$

The right side of the bracketed term in (6.17) is no longer a sum of squares and can easily be negative. Hence a single species that evolves by indirect effects can decrease community density. The fact that community density increases in the majority of cases (Fig. 6.1 and 6.2) is an emergent property, found only when an entire community of interacting organisms are considered together.

To summarize, these models represent a mathematically simple but biologically unrealistic community with random interdependencies among its members. When every member is allowed to change its influence on others according to the rules of indirect effects, some are enhanced and others inhibited, but the community as a whole increases its density in 98 of the 100 simulation runs conducted. No member is directly sensitive to community density, but community density nonetheless increases.

Next it is necessary to show that the same conclusion holds for a more realistic set of models. Three steps toward biological realism will be taken in this chapter: (1) imposing "carrying capacities"; (2) decreasing the connectivity of the matrix, or the ability of a species to affect other species; and (3) diversifying the mathematics and attaching a physical interpretation to species effects in terms of community structure and function.

Carrying Capacities

The most obvious defect in the previous models is the absence of external constraints on density. Given sufficient iterations of the equations, the populations become astronomically large. This problem is usually solved by employing the Lotka-Volterra equations

$$\frac{dN_i}{dt} = \frac{M_i N_i}{K_i} \left[K_i + \sum_{j=1}^{s} N_i a_{ij} \right] \quad \textbf{(6.18)}$$

where M_i, K_i, and N_i are the intrinsic rate of increase, the carrying capacity, and the density of species i; and a_{ij} is the per capita effect of species j on species i. Unfortunately, there is a strong reason why this classic formulation cannot be used in the following models. Mutualistic relationships, in which both a_{ij} and a_{ji} are positive, tend to make equilibrium densities unstable. The mutualists embark upon an orgy of benevolence, and their numbers expand to infinity. Needless to say, this is not a meaningful statement about nature, but rather a flaw in the Lotka-Volterra equations (May 1976).

A more realistic way to model mutualisms and many other interactions is as follows (May 1976):

$$\frac{dN_i}{dt} = \frac{M_i N_i}{K_i}\left[K_i + \sum_{j=1}^{s} N_j a_{ij} / (L + N_j) - N_i\right] \tag{6.19}$$

where L is a constant. Here the per capita effect of species j on species i is represented by $a_{ij}/(L + N_j)$. The total effect of species j on species i is asymptotic and can never exceed the value of a_{ij}. For example, an earthworm may increase plant productivity, but only to a certain higher level. Eventually a new limiting factor will always impose itself. Equation (6.19) is well suited for the investigation of indirect effects and forms the nucleus for the following models. However, before continuing, I should stress that these community-matrix models differ from others—not only in the equations used, but also in the questions asked. In particular, I am not concerned with the evolution of stability and its relation to species diversity (May 1975, Goodman 1975, McNaughton 1977). Rather, I am focusing on the evolution of increased community productivity, in which the extinction of undesirable species is a vital part of the process. A functional approach to communities may eventually contribute something to the diversity-stability controversy. Stability is likely to be a prerequisite for productivity and may be enhanced by diversity in some cases—but only if the relationships between species are rigidly defined. An exploration of the problem is beyond the scope of this chapter.

Consider a community composed of s niches, filled by s species. If the niches are completely autonomous from each other, then at equilibrium the species' densities will be

$$\begin{bmatrix} N_1 \\ N_2 \\ \cdot \\ \cdot \\ \cdot \\ N_s \end{bmatrix} = \begin{bmatrix} K_1 \\ K_2 \\ \cdot \\ \cdot \\ \cdot \\ K_s \end{bmatrix} \tag{6.20}$$

where N_i and K_i are the density and carrying capacity, respectively, of species i. Now consider the more interesting situation in which the niches are not autonomous. Instead the carrying capacities are to some extent a function of the surrounding community. Hermit crabs' carrying capacities depend on shells. Ground birds' carrying capacities depend on cover. Nematodes' carrying capacities depend on the proper size of soil spaces. Many species' carrying capacities can be completely eclipsed by a toxin produced by another species. Other examples of biotically determined carrying capacities are provided in Chapter 5.

From (6.19), the new carrying capacities may be written (for species 1):

$$K'_1 = K_1 + a_{12}N_2/(L + N_2) + a_{13}N_3/(L + N_3) \cdots a_{1s}N_s/(L + N_s) \tag{6.21}$$

At equilibrium, the whole community may be expressed:

$$\begin{bmatrix} N_1 \\ N_2 \\ \cdot \\ \cdot \\ \cdot \\ N_s \end{bmatrix} \begin{bmatrix} K_1 \\ K_2 \\ \cdot \\ \cdot \\ \cdot \\ K_s \end{bmatrix} + \begin{bmatrix} 0 & a_{12} & a_{13} & \cdots & a_{1s} \\ a_{21} & 0 & a_{23} & \cdots & a_{2s} \\ \cdot & & & & \\ \cdot & & & & \\ \cdot & & & & \\ a_{s1} & & & & 0 \end{bmatrix} \begin{bmatrix} N_1/(L + N_1) \\ N_2/(L + N_2) \\ \cdot \\ \cdot \\ \cdot \\ N_s/(L + N_s) \end{bmatrix} \tag{6.22}$$

To explore evolution in this model, the following computer simulation was utilized:

1. Random values for a_{ij}, uniformly distributed between $+1$ and -1, are entered into the matrix. The sum of the absolute values of each column is then normalized to a value of W. Thus, each species has a fixed adaptive potential of magnitude W that can be distributed in any way, positively or negatively, to the rest of the community.
2. Arbitrary numbers are entered for starting densities, and the equations iterated until equilibrium densities are obtained. The equations may have multiple equilibrium solutions; however, the simulation operates only on the first equilibrium encountered with this procedure. Negative species densities are set to 0.
3. In preceding models, the effect of species j on i equaled $a_{ij}N_j$. However, because this is an equilibrium model, species effects occur through many pathways

(e.g., i's effect on j, through its effect on k, and so on). To determine the true implications of changing an element, each element is successively incremented by a small value (v) and the equations iterated until they come to a new equilibrium. This procedure is used to generate two new matrices:

a. a "species welfare" matrix (**B**). Every element b_{ij} in this matrix gives the effect of incrementing the corresponding element a_{ij} of the **A**-matrix on the density of the species j.

b. a "community welfare" matrix (**C**). Every element c_{ij} in this matrix gives the effect of incrementing the corresponding element a_{ij} of the **A**-matrix on the density of the entire community.

4. Step 3 shows how changing an element affects both the individual species and the entire community. Next the elements of the **A**-matrix are changed according to the rules of indirect effects

$$a'_{ij} = a_{ij} + z\,|a_{ij}|\,b_{ij} \qquad \textbf{(6.23)}$$

where $|a_{ij}|$ is the absolute value of a_{ij} and z is a constant, chosen such that a'_{ij} is only slightly different than a_{ij} (a requirement of gradual evolution). When every element has been similarly changed and the columns renormalized to a value of W, those elements that had the greatest effect on species welfare will have proportionately changed the most. As in the preceding model, every species is sensitive only to its own welfare (the **B**-matrix). The community welfare matrix is generated for comparative purposes, and is not used in the evolution of the **A**-matrix.

5. After a sufficient number of iterations, the matrix elements no longer change. The community has reached an "evolutionary equilibrium" in which each species has optimally distributed its adaptive potential toward the rest of the community, to maximize its own abundance through indirect effects.

We want to know if the total community welfare is also maximized at this point. If so, then the evolution of indirect effects has caused the community to reach the top of an "adaptive peak" in terms of community welfare. This may be determined by changing the formula by which the elements evolve to

$$a'_{ij} = a_{ij} + z\,|a_{ij}|\,c_{ij} \qquad (6.24)$$

Now each species is directly sensitive to community welfare (the **C**-matrix). If evolution by equation (6.24) yields higher community densities than evolution by equation (6.23), then indirect effects cannot be said to have maximized community fitness.

Results. The relevant parameters and the values used in the simulation runs are summarized in Table 6.1. Ideally, two parameters (v and z) should be as small as possible to remain biologically realistic. The values used are the smallest that allowed evolutionary equilibrium to be reached in a practical amount of computer time. In the following analysis, two parameters have been varied to explore their effects on community evolution: (1) s, or the number of species initially present in the community, and (2) K/W, the size of the abiotically determined carrying capacity relative to the adaptive potential of one species to alter the densities of other species in the community. A total of 273 simulation runs were conducted, whose outcomes fell into three categories. These categories will first be discussed, and then their relative proportions as a function of s and K/W.

Category 1: Community Welfare Maximized. Table 6.2 presents a single example in detail. At $t = 0$, the **A**-matrix elements are simply random numbers, with the sum of the absolute values of each column adding to 40 ($W = 40$). The species densities, which total 58.26, are the equilibrium densities for that set of element values, determined by successive iterations of equation (6.21).

The species-welfare matrix gives the effect of a species on itself by incrementing its column elements. Thus, if species 1 changes its

TABLE 6.1 Parameters of the carrying-capacity model, and values used in the simulation runs. Ideally, v and z should be as small as possible. The values used represent a compromise: Allowing a reasonably gradual evolution of the elements, while assuring that an evolutionary equilibrium is reached within 50 to 100 iterations.

Parameter	*Definition*	*Value(s)*
s	Initial species number	3–7
K	Abiotically determined carrying capacity	5–60
W	Potential of each species to affect the carrying capacity of others (adaptive potential)	40
v	Element increment used to generate **B** and **C** matrices	1
z	Coefficient governing rate of evolution	1

TABLE 6.2 A simulation run of model in which community density is maximized. For this run, $K = 15$ and $W = 40$. Columns for extinct species are set to zero for visual convenience. See text for explanation.

t	0					10					50				
Species	1	2	3	4	5	1	2	3	4	5	1	2	3	4	5
A-matrix	0	−12.8	12.3	7.9	−11.3	0	−27.5	33.3	1.0	0	0	−.032	40	0	0
	−11.5	0	11.7	4.5	−20.5	−28.8	0	.284	34.7	0	−.031	0	0	40	0
	11.0	−10.5	0	−16.4	−4.2	10.9	−.38	0	−2.04	0	39.9	0	0	0	0
	6.9	11.3	−1.2	0	3.8	−.034	11.5	−.884	0	0	0	39.9	0	0	0
	−10.3	−5.3	−14.6	11.0	0	−.181	−.559	−5.52	2.17	0	0	0	0	0	0
Species welfare	0	.019	.062	.085	−.184	0	−.274	.096	−.036	0	0	0	.098	0	0
	−.5	0	−.170	.227	.199	−.26	0	−.036	.102	0	0	0	0	.098	0
	.117	.209	0	.219	−.352	.272	−.098	0	−.020	0	.098	0	0	0	0
	.017	−.033	−.030	0	.091	−.085	.258	−.026	0	0	0	.098	0	0	0
	.037	−.134	.044	−.097	0	0	0	0	0	0	0	0	0	0	0
Community welfare	0	.377	.383	.861	.634	0	.579	.590	.602	0	0	.941	.938	.941	0
	−.009	0	.002	.009	−.006	.594	0	.603	.618	0	.941	0	.941	.938	0
	.738	.359	0	.840	.622	.865	.871	0	.900	0	.938	.941	0	.941	0
	.579	.298	.303	0	.499	.860	.855	.870	0	0	.941	.938	.941	0	0
	.366	.196	.193	.461	0	0	0	0	0	0	0	0	0	0	0
Species density	14.6	4.4	4.5	24.1	10.4	19.8	19.9	20.6	22.0	0	48.0	48.0	48.0	48.0	0
Community density	58.26					82.46					192.35				

effect on species 2 from −11.5 to −10.5, its own density will decrease by 0.5 (element 2,1 in the species-welfare matrix). However, if species 2 changes its effect on species 1 from −12.80 to −11.80, its own density will increase by 0.019. This is true because even though species 1 has a negative direct effect on species 2, it has a net positive effect through its impact on species 3, 4, and 5.

The community-welfare matrix gives the effect of these same increments on the community at large. Thus, even though species 2 decreases its own density by changing its effect on species 5 from −5.3 to −4.3, it increases the density of the entire community by 0.196 (element 5,2 in the community-welfare matrix). In this particular case there is a conflict between the interests of species 2 and the interests of the whole community. By comparing the species-welfare matrix with the community-welfare matrix, it can be seen that many other conflicts exist. In fact, there is no close correspondence between the two matrices.

Each species then evolves to maximize its own density. Thus, species 1 makes its effect on species 2 more negative than it was before, and species 2 makes its effect on species 1 more positive, according to equation (6.21). After 10 iterations of this procedure, the evolving matrix is shown in Table 6.2 as $t = 10$. Species 5, which was initially the third most abundant species, has gone extinct because of its intensely negative effect on the community. Paired symbiotic interactions have deepened between species 1 and 3, 2 and 4. The relationship between species 1 and 2 also has changed. Species 1 was initially slightly beneficial to species 2, largely through its negative effect on species 5. Now that species 5 is extinct and species 1 has intensified its negative effect on species 2, the relationship between the two has become mutually antagonistic. The totality of these interactions has been positive. The community density has risen from 58.2 to 82.4.

By the fiftieth iteration (Table 6.2, $t = 50$), the community has neared its evolutionary equilibrium in which the matrix elements no longer change. Complete mutualism exists between species 1 and 3, 2 and 4. The antagonistic interactions between species 1 and 2, 3 and 4 have diminished because of the limited amount of adaptive potential available. Each species gains more by supporting its mutualist than by destroying its antagonist, and it cannot do both.

From the community-welfare matrix, it may be seen that incrementing any element provides equal benefit to the community. From the species-welfare matrix, it is obvious that each species only benefits from incrementing its effect on its mutualist; but since this does not conflict with community welfare in any way, the community density of 192.35 represents the top of an adaptive peak. Making the species directly sensitive to community welfare (equation 6.22) does not increase the community density further.

Notice, however, that this adaptive peak represents only a local optimum. Had the species been sensitive to community welfare from the beginning ($t = 0$), it would have evolved into a five-species mutualistic network, with an even higher community density. Nevertheless, it may be said for this example that a community whose members are motivated entirely by self-interest has evolved into a totally mutualistic, cooperating system. Furthermore, by allowing other species with different properties to recolonize niches vacated by extinct species, one could easily obtain a globally optimal community density by any standard. A totally mutualistic community does not exclude predator–prey relationships, since consumers may increase the productivity of their resources (Chapter 3).

Category 2: Community Welfare Increased but not Maximized. In some simulation runs, the evolution of indirect effects increased community density, but did not maximize it. One example is shown in Table 6.3. The **A**-matrix for this run happened to receive a large number of negative elements, such that species 4 goes extinct immediately and species 3 is rare at equilibrium. There are only two mutualistic interactions: a strong one between species 2 and 5, and a weak one between species 1 and 2.

By the 20th iteration, species 3 has been driven to extinction (it actually went extinct almost immediately) and the mutualistic relationship between species 2 and 5 has intensified. The community density has risen substantially from 56.3 to 110.99. Species 1 and 2 have a weak mutualistic relationship, but species 2 finds it more advantageous to support species 5. At the evolutionary equilibrium ($t = 50$), species 1 has become autonomous. Nothing it does can feed back either positively or negatively to itself. The equilibrium is far below the adaptive peak for community welfare, in which all three species would engage in a mutualistic network.

Category 3: Community Welfare Decreased. In some simulation runs, the evolution of indirect effects actually caused the community density to decrease, as shown for one example in Table 6.4. The **A**-matrix chanced to receive a very large number of negative elements, such that species 3 and 5 immediately go extinct and there are no opportunities for mutualistic relationships. By the 10th reiteration, species 4 has departed. Species 2 has lessened its negative effect on species 1, but so also has 1 lessened its positive effect on 2. Because the rate of evolution depends on the intensity of the interaction [equation (6.21)], the relationship eventually becomes

TABLE 6.3 A simulation run of model in which community density increases but is not maximized. For this run, $K = 15$ and $W = 40$. Columns for extinct species are set to 0 for visual convenience. See text for explanation.

t	0					20					50				
Species	1	2	3	4	5	1	2	3	4	5	1	2	3	4	5
A-matrix	0	−8.58	−9.32	0	15.5	0	.054	0	0	.194	0	.001	0	0	0
	2.97	0	−3.74	0	9.00	.327	0	0	0	38.6	3.28	0	0	0	40
	.804	−9.71	0	0	−11.9	−1.24	−.68	0	0	−1.12	−1.24	0	0	0	0
	−27.1	−9.12	10.1	0	−3.53	−34.8	−.26	0	0	.077	−34.8	0	0	0	0
	−9.10	12.5	−16.7	0	0	−.649	39.0	0	0	0	−.648	40	0	0	0
Species welfare	0	.003	.005	0	−.097	0	.041	0	0	−.004	0	.042	0	0	−.004
	.062	0	−.002	0	.198	0	0	0	0	.093	0	0	0	0	.092
	−.734	−.486	0	0	−1.10	0	0	0	0	0	0	0	0	0	0
	0	0	0	0	0	0	0	0	0	0	0	0	0	0	0
	.302	.179	−.036	0	0	0	.099	0	0	0	0	.099	0	0	0
Community welfare	0	.556	.109	0	.497	0	.867	0	0	.862	0	.872	0	0	.868
	.860	0	.185	0	.830	.68	0	0	0	.929	.675	0	0	0	.93
	−1.21	−1.74	0	0	−1.54	0	0	0	0	0	0	0	0	0	0
	0	0	0	0	0	0	0	0	0	0	0	0	0	0	0
	1.13	1.20	.244	0	0	.685	.94	0	0	0	.681	.944	0	0	0
Species density	17.3	21.8	1.5	0	15.6	15.2	48.8	0	0	46.9	15.0	50.0	0	0	47.95
Community density	56.33					110.99					113.01				

TABLE 6.4 A simulation run of model in which community density decreases. For this run, $K = 15$ and $W = 40$. Columns for extinct species are set to 0 for visual convenience. See text for explanation.

t	0					10					50				
Species	1	2	3	4	5	1	2	3	4	5	1	2	3	4	5
A-matrix	0	−6.88	0	.485	0	0	−2.58	0	0	0	0	−1.71	0	0	0
	2.45	0	0	−17.7	0	.820	0	0	0	0	−.103	0	0	0	0
	−31.7	−4.57	0	−1.36	0	−34.4	−3.36	0	0	0	−35.1	−3.43	0	0	0
	−5.08	−12.8	0	0	0	−3.87	−22.4	0	0	0	−3.94	−22.9	0	0	0
	−.764	−15.7	0	−20.4	0	−.831	−11.5	0	0	0	−.85	−11.8	0	0	0
Species welfare	0	.079	0	−.057	0	0	.009	0	0	0	0	−.001	0	0	0
	−.120	0	0	−.152	0	−.022	0	0	0	0	−.017	0	0	0	0
	0	0	0	0	0	0	0	0	0	0	0	0	0	0	0
	.100	−.463	0	0	0	0	0	0	0	0	0	0	0	0	0
	0	0	0	0	0	0	0	0	0	0	0	0	0	0	0
Community welfare	0	.476	0	.361	0	0	.616	0	0	0	0	.598	0	0	0
	.353	0	0	.237	0	.550	0	0	0	0	.567	0	0	0	0
	0	0	0	0	0	0	0	0	0	0	0	0	0	0	0
	.291	.273	0	0	0	0	0	0	0	0	0	0	0	0	0
	0	0	0	0	0	0	0	0	0	0	0	0	0	0	0
Species density	11.7	9.7	0	5.9	0	13.4	15.5	0	0	0	13.6	14.9			
Community density	27.44					28.89					28.91				

mutually antagonistic ($t = 50$), from which densities can only spiral downward.

Now we are ready to explore community evolution in the context of the general model. In Chapter 3 it was stated intuitively that for indirect effects to operate successfully, a community must be "strong" relative to any single member. One way to simulate this is by varying the ratio of K/W, where K is the abiotic component of the carrying capacity and W is the potential for one species to affect the carrying capacity of others in its community. Figure 6.4 displays the proportion of optimal solutions (category 1) and the proportion

FIGURE 6.4 The proportion of simulation runs yielding optimal community density (solid line) and decreased community density (dashed line) as K is varied relative to W.

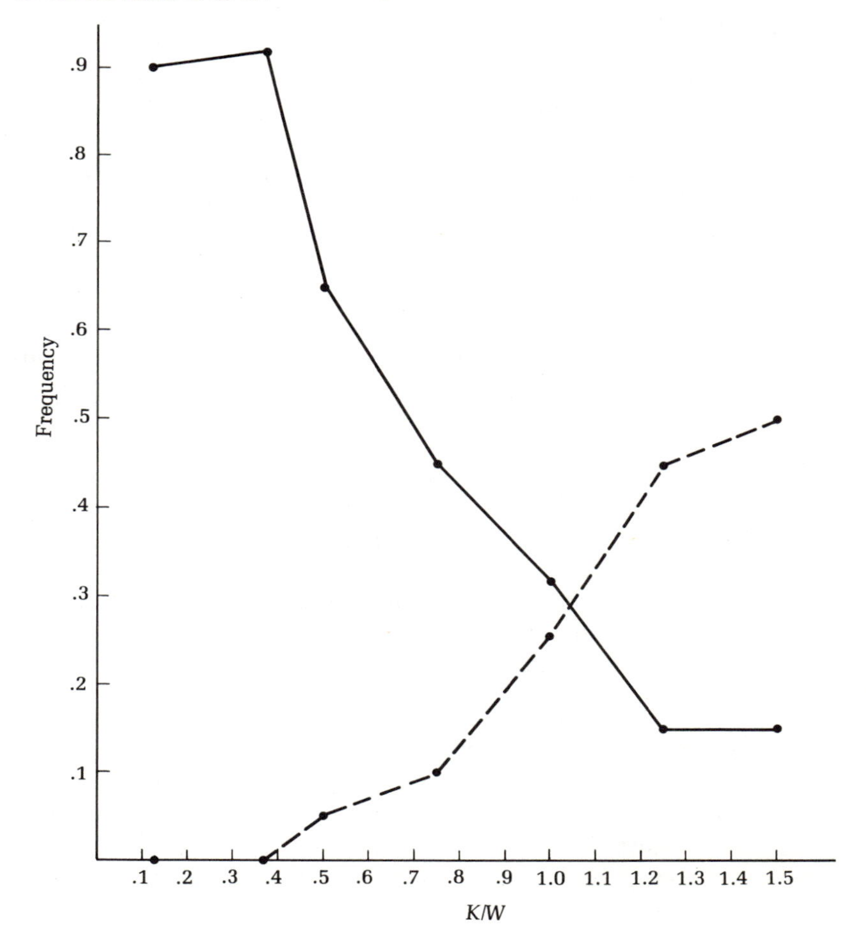

of solutions yielding decreased community density (category 3) for five-species communities as a function of *K/W*. The figure summarizes the results of 170 simulation runs, in which *W* was held constant at 40 and *K* varied from 5 to 60.

At low values of *K/W*, each species can easily drive another extinct. Approximately 90% of all simulation runs reach an adaptive peak in terms of community density, and in no case does evolution by indirect effects decrease community density.

At intermediate values of *K/W*, the proportion of optimal solutions declines rather sharply. Antagonistic interactions between two species still end in an extinction, but it takes a longer time to accomplish. By the time it occurs, the rest of the community is well on its way toward forming a mutualistic network, and the surviving member of the negative interaction is likely to find itself autonomous as in the example provided in Table 6.3. However, the proportion of simulation runs leading to a decrease in community density is still low.

At high values of *K/W*, a species cannot drive another extinct, even with the full weight of its adaptive potential. The proportion of optimal solutions has sunk very low, and a decrease in community density occurs in 50% of the simulation runs. This result is what one might expect in a community with random initial interdependencies and little potential for evolution by indirect effects.

Figure 6.5 displays the effect of species number for $K = 15$ and $W = 40$. At least within the range of $s = 3$ to 7, species number seems to have no effect on the proportion of optimal outcomes. Unfortunately, the program consumes too much computer time to generate adequate sample sizes for large communities. However, those few runs I have conducted appear to conform to the pattern.

We may summarize by saying that the imposition of carrying capacities does not fundamentally alter the conclusion of the simple linear model, given a sufficiently large control of the community over its members. The carrying-capacity model also suggests that in addition to causing an increase in community density, evolution by indirect effects could actually maximize it under the proper circumstances.

Constraints on Adaptive Potential

The preceding models assume that every species has the potential to evolve positive or negative effects toward any other species in its community. However, in nature many species will at times be simply unable to evolve in certain directions, even though it would benefit them to do so. How sensitive is community evolution to

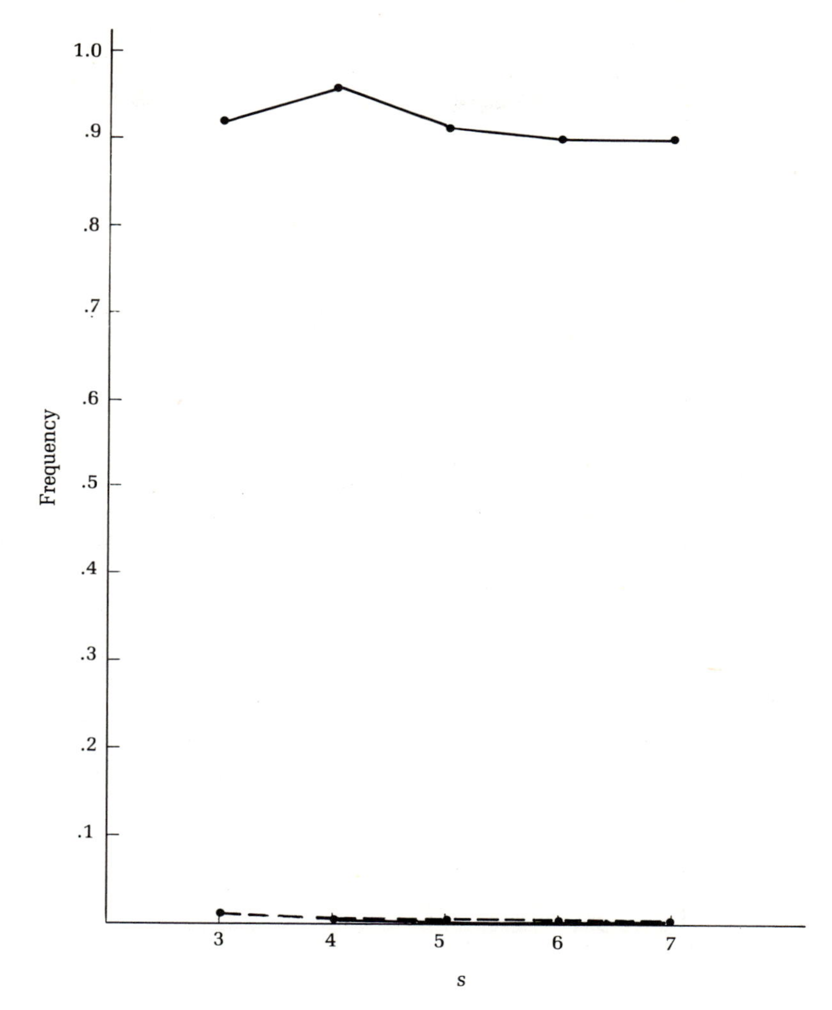

FIGURE 6.5 The proportion of simulation runs yielding optimal community density (solid line) and decreased community density (dashed line) as a function of species number (s).

constraints on the evolution of indirect effects? This question can be approached by a simple modification of the carrying-capacity model. After assigning random numbers to the **A**-matrix and normalizing the columns to a value of W, a certain number of elements are chosen at random and designated "nonevolvable." As other elements change according to the rules of indirect effects, the nonevolvable elements retain their starting values throughout the simulation run.

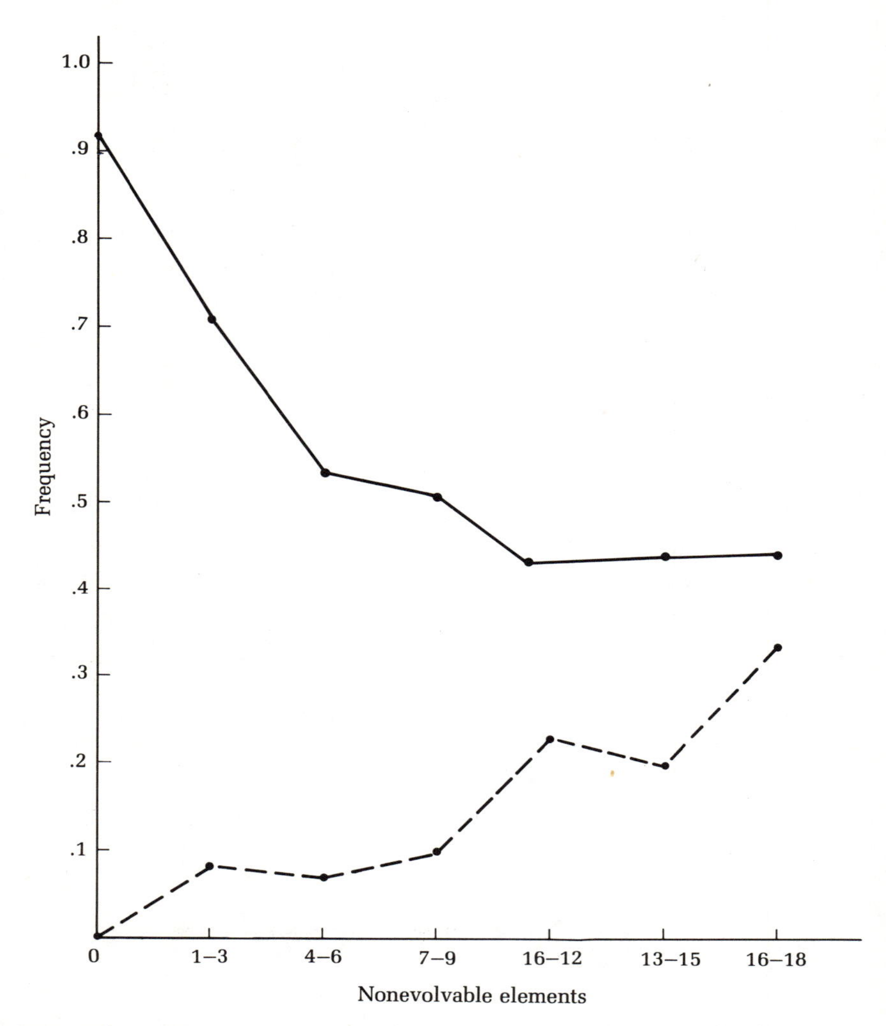

FIGURE 6.6 The proportion of simulation runs yielding optimal community density (solid line) and decreased community density (dashed line) as the number of nonevolvable elements in the matrix are increased.

Figure 6.6 displays the results of 210 simulation runs for five-species communities, in which $K = 15$ and $W = 40$. It shows the proportion of simulation runs yielding optimal (solid line) and decreased (dashed line) community density as a function of the number of nonevolvable elements in the matrix (there are 20 off-diagonal elements in a five-species community matrix). The proportion of optimal solutions is initially quite sensitive to an increase in nonevolvable elements, but then it levels off. The proportion of runs yielding a decreased density acts in the opposite fashion, ris-

TABLE 6.5 A simulation run of model 5.3 in which 12 of the 20 off-diagonal elements are nonevolvable (designated in boldface). See text for explanation.

t	0					5					45				
Species	1	2	3	4	5	1	2	3	4	5	1	2	3	4	5
A-matrix	0	17.8	**−21.8**	**1.42**	**15.2**	0	15.9	**−21.8**	**1.42**	**15.2**	0	2.55	**−21.8**	**1.42**	**15.2**
	−5.79	0	−1.19	**−10.3**	6.82	**−5.79**	0	−1.95	**−10.3**	1.05	**−5.79**	0	−4.02	**−10.3**	.317
	8.90	−9.91	0	3.50	**12.2**	**8.90**	−11.8	0	3.15	**12.2**	**8.90**	−25.2	0	.400	**12.2**
	−12.2	**−7.46**	2.83	0	−5.82	**−12.2**	**−7.46**	2.07	0	11.59	**−12.2**	**−7.46**	0	0	12.33
	13.0	**4.83**	**−14.2**	24.8	0	**13.0**	**4.83**	**−14.2**	25.12	0	**13.0**	**4.83**	**−14.2**	27.8	0
Species welfare	0	.023	0	0	0	0	−.018	0	0	0	0	−.02	0	0	0
	0	0	−.18	0	−.137	0	0	−.353	0	.171	0	0	−1.18	0	.281
	0	−.034	0	.004	0	0	−.002	0	.022	0	0	−.004	0	.027	0
	0	0	.473	0	1.26	0	0	.118	0	.260	0	0	.207	0	.269
	0	0	0	−.007	0	0	0	0	.041	0	0	0	0	.042	0
Community welfare	0	.456	0	0	0	0	.309	0	0	0	0	.238	0	0	0
	0	0	.566	0	.525	0	0	.861	0	.856	0	0	−.175	0	−.049
	0	.449	0	.052	0	0	.226	0	.442	0	0	.143	0	.397	0
	0	0	2.03	0	1.96	0	0	.998	0	1.01	0	0	1.10	0	1.18
	0	0	0	.074	0	0	0	0	.763	0	0	0	0	.774	0
Species density	20.0	14.0	23.2	.714	18.3	16.2	4.53	27.8	15.2	29.3	12.6	2.86	23.7	15.8	30.4
Community density	76.33					93.15					85.43				

ing steeply only as large numbers of nonevolvable elements are obtained. Notice the similarity between Figure 6.6 and Figure 6.4.

Table 6.5 provides a single example in detail. Twelve of the 20 off-diagonal elements are nonevolvable in this run, designated by italics in the **A**-matrix (notice that the corresponding elements of the **B**- and **C**-matrices are 0). The potential for change is severely limited. Species 1 cannot change its effect on any other species.

FIGURE 6.7 The proportion of simulation runs yielding optimal community density (solid line) and decreased community density (dashed line) as K is increased relative to W. An average of one-third of the matrix elements is nonevolvable.

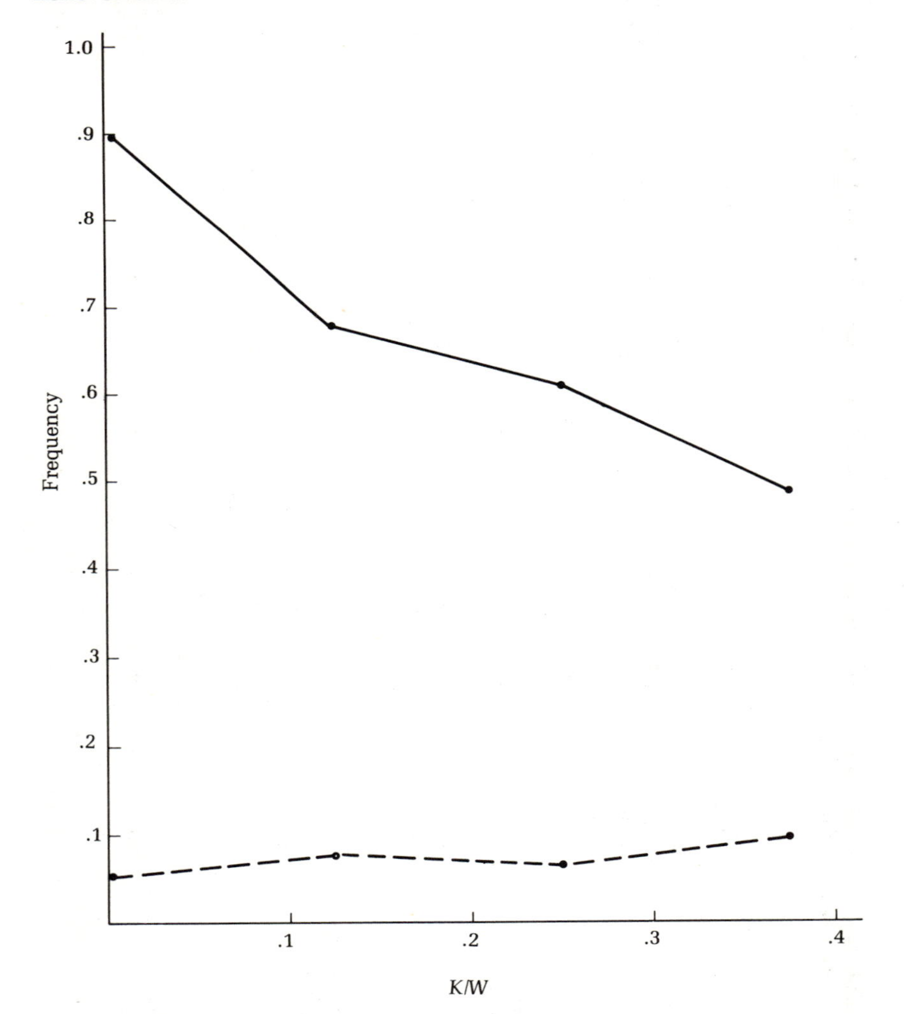

Species 3 cannot change its negative effect on its chief benefactor, species 5. Neither can species 4 inhibit its chief antagonist, species 1. By the 5th iteration, the evolving pathways have become clear. Species 4 and 5 are engaging in a mutualistic interaction. A possible mutualistic interaction between 3 and 4 is not forming because species 3 finds it more advantageous to antagonize species 2. Species 2 is evolving antagonisms toward both 1 and 3. By the 45th iteration, the community has neared its evolutionary equilibrium. The community density has increased only very slightly over its starting value, and is still far below any adaptive peaks.

This model indicates that the presence of nonevolvable elements can indeed impair the process of community evolution, but in what way? Figure 6.6 bears a strong resemblance to Figure 6.4, in which K, the abiotically determined carrying capacity, is increased relative to W. Perhaps nonevolvable elements produce the same effect as K because they reduce the amount of adaptive potential available to respond to indirect effects. For instance, in Table 6.5 the sum of the absolute value of the elements in column three equals 40, but only 4.02 of it is evolvable. Had it been more, species 3 could have driven species 2 to extinction and returned to a mutualistic relationship with species 4. To test this idea, 178 simulation runs were performed in which an average of one-third of the elements were designated nonevolvable, $W = 40$, and K was varied from 15 to 0.1. The results are shown in Figure 6.7. At the lowest value of K/W, the frequency of runs yielding optimal and decreased community density becomes virtually identical to the carrying-capacity model for $K = 15$. This is interesting because the term $W/3$, representing the average nonevolvable portion of a species' adaptive potential, equals 13.33. It therefore appears that given sufficient *evolvable* adaptive potential, the presence of nonevolvable elements does not greatly impair the process of community evolution. Even when a species on the average cannot respond to the effects of one-third of the other species in its community, a mutualistic network of evolvable elements emerges in the majority of cases.

Discussion

All the preceding simulations indicate that biological communities can evolve into superorganisms of sorts without any species being directly sensitive to community welfare. However, while the models incorporating carrying capacities and constraints on adaptive potential are improvements over the linear model, they are still poor caricatures of the real world. A long list of their deficiencies would include the following:

1. All the preceding models use density as a measure of species/community fitness. Density will not be a true measure for real communities, in which inverse relations exist between density and productivity (see Chapter 3). The problem is further complicated by between species differences in size, generation time, and adaptive potential—none of which have been considered in the models.
2. In nature, many powerful effects occur through multilink pathways. Species *i* may affect *j* by affecting *k*, and so on. Multilink effects are not absent in the models, but they are weak relative to single-link effects. Hence, the very important process of a species influencing another through an intermediary is inadequately represented.
3. In all three models, only one kind of interaction is represented—the effect of a species on the density of another. This ignores a great diversity of relationships among organisms in nature. In particular, one important class of interactions concerns species *i* altering the *effect* of species *j* on species *k*, a modification of matrix element a_{kj} as opposed to N_j (see p. 112 for an example). In this way, the community can control not only the density but also the column elements of its members.
4. The idea of a species depending upon several members of the community at once is missing. In most simulation runs, a species tended to concentrate its adaptive potential either for or against one other member of the community; yet in nature, such extreme specialization is usually not possible.
5. Many of the positive but suboptimal outcomes of simulation runs, such as that represented in Table 6.3, are artificial and probably have limited meaning in nature. It is unlikely that a species can ever become truly autonomous in a community.
6. The concept of a fixed adaptive potential is artificial, particularly when nonevolvable elements are included. While there is little doubt that limitations do exist, the underlying constraints are likely to be far more complex.
7. Most important, the interactions between species in

> nature are fundamentally tied to ecosystem structure and function. A species does not simply benefit another—it benefits it by supplying a needed resource, providing protection against a predator, removing a waste product, improving the microclimate, and so on. These activities will, in turn, define relationships with other species. In short, many of the constraints on the evolutionary potential of species in a community are defined by the environment, yet in most evolutionary models the environment of a species is woefully underrepresented by a single "carrying capacity" term. The concept of a carrying capacity serves a valuable function in population ecology as a black box, but any realistic model of community evolution must open that box and closely examine the physiology of biological communities.

It is far beyond the scope of this chapter to make these modifications here. However, I would like to illustrate some of them with a single example. Figure 6.8 displays an imaginary eight-species community centering around a plant and a microbe that, let us say, benefits the plant by converting phosphorus from a nonusable to a usable form. Solid and dashed arrows designate evolvable and nonevolvable effects respectively.

The plant is destructively attacked by a herbivorous insect. If the insect had any intraspecific interference mechanisms, it could control its own activities, but it does not. It is an exploitation competitor and so reduces the plant density to a suboptimal level.

The plant responds by producing a toxin in its leaves. The toxin successfully controls the insect, but it has an undesirable side effect. When the leaves fall off the plant and enter the soil, the toxin leaches out and inhibits the microbe.

This situation can be remedied by a fourth species that detoxifies the toxin after it has entered the soil. However, detoxification takes energy, which makes species 4 a poor competitor for resource *A*, which instead flows to a fifth species, competitor *A*. But competitor *A*'s superiority may be inhibited by a sixth species, inhibitor *A*, which derives its energy in a nondestructive fashion from the plant. A similar situation exists between the microbe, competitor *B*, and inhibitor *B*.

The equations embodying these relationships, parameter values, and ranges for the evolvable coefficients are provided in Table 6.6. Species densities before and after evolution are shown in Table 6.7. The fact that community density increases is not surprising

TABLE 6.6 Equations for equilibrium densities and parameter values for model. For the evolvable parameters, the values initially entered into the program were the lower ends of the ranges.

Species	*Description*	*Equation*
1	Plant	$N_1 = K_1 + a_{12}(N_1/(L_1 + N_1)) - a_{13}N_3$
2	Microbe	$N_2 = K_2 + (1 - a_1/(L_2 + a_1))K_3(a_{78}N_8/(L_3 + N_8))$
3	Insect	$N_3 = N_1a_{31}/a_2$
4	Detoxifier	$N_4 = K_2 + K_3(a_{56}N_6/(L_3 + N_6))$
5	Competitor *A*	$N_5 = K_2 + K_3(1 - a_{56}N_6/(L_3 + N_6))$
6	Inhibitor *A*	$N_6 = a_{61}N_1$
7	Competitor *B*	$N_7 = K_2 + (1 - a_1/(L_2 + a_1))K_3(1 - a_{78}N_8/(L_3 + N_8))$
8	Inhibitor *B*	$N_8 = a_{81}N_1$
a_1	Below ground toxin	$a_1 = a_2(1 - a_3N_4/(L_3 + N_4))$

Evolvable parameters	*Description*	*Evolvable range*
a_2	Toxin produced by plant	1–10
a_3	Efficiency of detoxifier	0–1
a_{12}	Effect of microbe on plant	30–50
a_{56}	Effect of inhibitor *A* on competitor *A*	0–1
a_{78}	Effect of inhibitor *B* on competitor *B*	0–1

Nonevolvable parameters	*Description*	*Value*
K_1	Abiotically determined carrying capacity for plant	50
K_2	Abiotically determined carrying capacity for microbe, detoxifier, competitor *A*, competitor *B*	1
K_3	Carrying capacity from resource *A* and *B*	40
L_1	Constant governing rate of asymptote for plant	15
L_2	Constant governing rate of asymptote for effect	3
L_3	Constant governing rate of asymptote for effect of inhibitors *A*, *B* on competitors *A*, *B*	15
a_{13}	Effect of insect on plant	1
a_{31}	Effect of plant on insect	.9
a_{61}	Effect of plant on inhibitor *A*	1
a_{81}	Effect of plant on inhibitor *B*	1

because that is the way the equations were built. This model lacks the virtue of the previous models in this chapter in that the relationships between species are not randomly constructed and therefore can say nothing about the intrinsic tendency of community evolution. Nevertheless, as an illustration, it contains several interesting features lacking in the random models:

TABLE 6.7 Species densities before and after evolution by indirect effects. The insect, competitor *A*, and competitor *B* have been inhibited. The others have been enhanced.

Species	*Description*	*Starting density*	*Final density*
1	Plant	27	73
2	Microbe	1	22
3	Insect	24	7
4	Detoxifier	1	38
5	Competitor *A*	41	4
6	Inhibitor *A*	27	73
7	Competitor *B*	31	5
8	Inhibitor *B*	27	73

1. The plant has the power to change its relationship with only one other member of the community—the herbivore (via toxin production). Yet its true adaptive potential is in fact much greater because two other species are dependent on the plant (the inhibitors) and will evolve in whatever way they can to improve its fitness. The true adaptive potential of the plant, therefore, consists of its own potential and that of its dependents. This process is lacking in the other models because of the weakness of multilink effects.

2. The inclusion of a species' dependents in calculating its adaptive potential carries an important corollary because the number of dependents a species has depends entirely on the function it performs in its community. A positive correlation, therefore, exists between community function and adaptive potential (W), which gives

FIGURE 6.8 An eight-species community centering around a plant and a microbe. Arrows give the effect of the species on one another. Solid and dashed lines indicate evolvable and nonevolvable effects respectively.

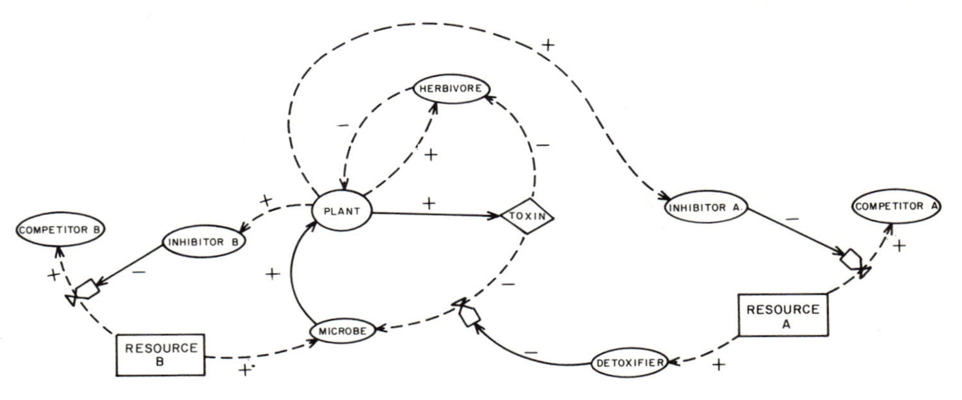

an added evolutionary advantage to species that benefit the community. As an example, in Figure 6.8 the plant possesses two dependents, while competitor *A* has none—a fact that makes all the difference in the outcome of community evolution. This is partially just the way the model is constructed, but it also has an element of biological realism in it: To have dependents, competitor A would have to do something for them, in which case it would have acquired a positive function in the community, at least for them. When dependents are considered, community function and adaptive potential become linked.

3. As far as the plant (and the community) is concerned, the toxin has both a positive and a negative effect. It controls the insect, but also inhibits the microbe. By reducing the rate of toxin production experienced by the microbe (but not the herbivore), the detoxifier eliminates the negative effect while retaining the positive. Such a subtle modification would have been impossible if community evolution operated only through the pathway of increasing or decreasing plant density.

4. Neither the microbe nor the detoxifier depend directly on the plant, but rather on the autonomous resources, *A* and *B*. At the beginning of the simulation run, neither evolved increases in their "community function." However, eventually positive feedback loops (helping the plant to help the inhibitor) strengthened to the point where it became advantageous to do so. The welfare of the microbe and the detoxifier became linked to that of the plant.

What do the models of this chapter say about the tendency of real biological communities to evolve into mutualistic networks? To me, their main virtue is their simplicity. One can be sure that a model requiring a large number of parameters, carefully balanced with each other, will not represent all the varied systems in nature. I therefore find it encouraging that the basic pattern of increasing community fitness can be produced by a linear model as simple as the one that began the chapter, and is robust against the various modifications to which the model was subjected. Perhaps it will still be robust when tested against truly realistic evolutionary models of biological communities, when they are built. Until then, I hope I have demonstrated the possibility that a functional approach to biological communities can be reconciled with evolutionary theory.

Literature Cited

Alexander, R. D. 1974. The evolution of social behavior. *Ann. Rev. Ecol. Syst.* 5:325–83.

Angerilli, N. P. D. C. and Beirne, B. P. 1974. Influences of some fresh water plants on the development and survival of mosquito larvae in British Columbia. *Can. J. Zool.* 52:813–18.

Archibald, C. P. 1975. Experimental observations of the effects of predation by goldfish on the zooplankton of a small saline lake. *J. Fish. Res. Bd. Can.* 32(9):1589–94.

Atsalt, P. R. and O'Dowd, D. J. 1976. Plant defence guilds. *Science* 193:24–29.

Attionu, R. J. 1976. Some effects of water lettuce (*Pistia stratiotes* L.) on its habitat. *Hydrobiologia* 50:245–54.

Ayala, F. J. 1968. Evolution of Fitness. II. Correlated effects of natural selection on the productivity and size of experimental populations of *Drosophila serrata. Evolution* 22:55–65.

Ayala, F. J. and Campbell, C. A. 1974. Frequency dependent selection. *Ann. Rev. Ecol. Syst.* 5:115–39.

Backiel, T. and LeCren, E. D. 1967. Some density relationships for fish—population parameters. In *The biological basis of freshwater fish production*, ed. S. D. Gerking, pp. 261–95. New York: Wiley.

Baker, K. F. and Cook, R. J. 1974. *Biological control of plant pathogens.* San Francisco: W. H. Freeman and Co.

Balduf, W. J. 1935. *Bionomics of Entomophagous Coleoptera.* St. Louis: J. S. Swift Co.

Barbosa, P. and Peters, T. M. 1970. The manifestations of overcrowding. *Bull. Ent. Soc. Am.* 16(2):89–93.

Barbosa, P.; Peters, T. M.; and Greenough, N. C. 1972. Overcrowding of mosquito populations: responses of larval *Aedes aegypti* to stress. *Envir. Ent.* 1:89–93.

Barsdate, R. H.; Prentki, R. T.; and Fenchel, T. 1974. Phosphorus cycle of model ecosystems: significance for decomposer food chains and effect of bacterial grazers. *Oikos* 25:239–51.

Batra, L. R. 1966. Ambrosia fungi: extent of specificity to ambrosia beetles. *Science* 153: 193–95.

———. 1972 Ectosymbiosis between ambrosia fungi and beetles, *Indian J. Mycol. Plant Pathol.* 2(2):165–69.

Batra, S. W. T. and Batra, L. R. 1967. The fungus gardens of insects. *Sci. Am.* 112–20.

Batzli, G. O. 1975. The role of small mammals in arctic ecosystems. In *Small mammals, their productivity and population dynamics*, IBP Handbook No. 5; eds. F. B. Golley, K. Petrusewucz, and L. Ryszkowski, pp. 243–268.

Bentley, B. L. 1976. Plants bearing extrafloral nectaries and the associated ant community: interhabitat differences in the reduction of herbivore damage. *Ecology* 57:815–20.

Bernstein, J. S. and Gordon, T. O. 1974. The function of aggression in primate societies. *Am. Sci.* 62:303–11.

Bethel, W. M. and Holmes, J. C. 1977. Increased vulnerability of amphipods to predation owing to altered behavior induced by larval acanthocophalans. *Can. J. Zool.* 55:110–28.

Binns, E. S. 1972. *Arctoseius cetratus* (Sellnick) (Acarina, Ascidae) phoretic on mushroom sciarid flies. *Acarologia* 14:350–56.

———. 1973. *Digamasellus fallax* leitner (Mesostigmata, Digamasellidae) phoretic on mushroom sciarid flies. *Acarologia* 15:10–17.

Boorman, S. A. and Levitt, P. R. 1973. Group selection on the boundary of a stable population. *Theor. Pop. Biol.* 4:85–128.

Boothe, P. N. and Knauer, G. A. 1972. The possible importance of fecal material in the biological amplification of trace and heavy metals. *Limn. & Oceanog.* 17:270–75.

Borchers, H. 1968. Distribution and composition of populations of mites symbiotic on necrophilous beetles. Ph.D. dissertation, Entomology Dept., Iowa State University.

Borden, J. H. 1974. Aggregation pheromones in the Scolytidae. In *Pheromones*, ed. M. C. Birch, pp. 135–60. New York: American Elsevier.

Branch, G. M. 1975. Intraspecific competition in *Patella cochlean* Born. *J. An. Ecol.* 44:263–83.

Brock, T. D. 1967. Relationship between standing crop and primary productivity along a hot spring thermal gradient. *Ecology* 48:566–71.

Brown, J. L. 1969. Territorial behavior and population regulation in birds: a review and re-evaluation. *Wilson Bull.* 81:293–329.

______. 1978. Avian communal breeding systems. *Ann. Rev. Ecol. Syst.* 9:123–56.

Bryant, E. H. and Hall, A. E. 1975. The role of medium conditioning in the population dynamics of the housefly. *Res. Pop. Ecol.* 16:188–97.

Case, T. J. and Gilpin, M. E. 1974. Interference competition and niche theory. *Proc. Nat. Acad. Sci.* 71:3073–77.

Charlesworth, B. 1978. A note on the evolution of altruism in structured demes. *Am. Nat.,* 113:601–605.

Charnov, E. L. and Krebs, J. R. 1975. The evolution of alarm calls: altruism or manipulation? *Am. Nat.* 109:107–12.

Chew, R. M. 1974. Consumers as regulators of ecosystems: an alternative to energetics. *Ohio J. Sci.* 74:359–70.

Christiansen, F. B. and Feldman, M. W. 1975. Subdivided populations: a review of the one- and two-locus deterministic theory. *Theor. Pop. Biol.* 7:13–38.

Clausen, C. P. 1976. Phoresy among entomophagous insects. *Ann. Rev. Ent.* 21:343–68.

Cohen, D. and Eshel, I. 1976. On the founder effect and the evolution of altruistic traits. *Theor. Pop. Biol.* 10:276–302.

Collins, N. C. 1975. Population biology of a brine fly (Diptera: Ephydridae) in the presence of abundant algal food. *Ecology* 56:1139–48.

Collins, N. C.; Mitchell, R.; and Wiegert, R. G. 1976. Functional analysis of a thermal spring ecosystem, with an evaluation of the role of consumers. *Ecology* 57:1221–32.

Colwell, R. K. 1973. Competition and coexistence in a simple tropical community. *Am. Nat.* 107:737–60.

Colwell, R. K. and Fuentes, E. R. 1975. Experimental studies of the niche. *A. Rev. Ecol. Syst.* 6:281–310.

Connell, J. H. 1975. Some mechanisms producing structure in natural communities: a model and evidence from field experiments. In *Ecology and Evolution of Communities,* eds. M. L. Cody and J. M. Diamond, pp. 460–90. Cambridge: Harvard Univ. Press.

Cooper, D. C. 1973. Enhancement of net primary productivity by herbivore grazing in aquatic laboratory microcosms. *Limn. and Ocean.* 18:31–37.

Costa, M. 1969. The association between Mesostigmatic mites and Coprid Beetles. *Acarologia* 11:411–28.

Crocker, R. L. and Major, J. 1955. Soil development in relation to vegetation and surface age at Glacier Bay, Alaska. *J. Ecol.* 43:427–48.

Dale, H. M. and Gillespie, T. 1976. The influence of floating vascular plants on the diurnal fluctuations of temperature near the water surface in early spring. *Hydrobiologica* 49:245–56.

Dart, J. K. G. 1972. Echinoids, algal lawn and coral recolonization. *Nature* 239:50–51.

Darwin, C. 1859. The origin of species. Penguin edition. New York: Penguin, 1968.

Dawkins, R. 1976. The selfish gene. Oxford: Oxford Univ. Press.

Dayton, P. K. 1975. Experimental evaluation of ecological dominance in a rocky intertidal algal community. *Eco. Monogr.* 45:137–57.

Disney, R. H. L. 1970. Association between black flies and prawns with a discussion of the phoretic habit in simulids. *J. Anim. Ecol.* 40:83–92.

______. 1971. Two phoretic black flies and their associated mayfly host. *J. Ent. Ser. A. Gen. Ent.* 46:53–61.

Dunbar, M. J. 1960. The evolution of stability in marine environments: natural selection at the level of the ecosystem. *Am. Nat.* 94:129–36.

______. 1972. The ecosystem as a unit of natural selection. *Trans. Conn. Acad. Arts. Sci.* 44:114–30.

Dunsmore, J. D. 1963. Effect of the removal of an adult population of *Ostertagia* from sheep on concurrently existing arrested larvae. *Aust. Vet. J.* 39:459–63.

Dyer, M. I. and Bokhari, U. G. 1976. Plant-animal interactions: studies of the effects of grasshopper grazing on blue grama grass. *Ecology* 57:762–72.

Eaton, G. C. 1966. The social order of Japanese macaques. *Sci. Amer.* 235:96–108.

Ellison, L. 1960. Influence of grazing on plant succession of rangeland. *Bot. Rev.* 26:1–78.

Eltringham, S. K. 1974. Changes in the large mammal community of Mweya Peninsula, Rwenzori National Park, Uganda, following removal of hippopotamus. *J. Appl. Ecol.* 11:855–65.

Enright, J. T. 1977. Diurnal vertical migration: adaptive significance and timing. Part 1. Selective advantage: a metabolic model. *Limn. and Oceanogr.* 22:856–72.

Errington, P. L. 1956. Factors limiting higher vertebrate populations. *Science* 124:304–7.

Esch, G. W.; Gibbons, J. W.; and Bourque, J. E. 1975. An analysis of the relationship between stress and parasitism. *Am. Mid. Nat.* 86:160–68.

Eshel, I. 1972. On the neighbor effect and the evolution of altruistic traits. *Theor. Pop. Biol.* 3:258–77.

______. 1977. On the founder effect and the evolution of altruistic traits: an ecogenetical approach. *Theor. Pop. Biol.* 11:410–24.

Estes, J. A. and Palmisano, J. F. 1974. Sea otters: their role in structuring nearshore communities. *Science* 185:1058–60.

Farish, D. J. and Axtell, R. C. 1971. Phoresy redefined and examined in

Macrocheles muscaedomesticae (Acarina, Macrochelidae). *Acarologia* 13:16–29.

Feeny, P. 1975. Biochemical coevolution between plants and their insect herbivores. In *Coevolution of animals and plants,* eds. L. E. Gilbert & P. H. Raven, pp. 3–19. Austin: Austin Univ. Press.

Felsenstein, J. 1976. The theoretical population genetics of variable selection and migration. *Ann Rev. Genet.* 10:253–80.

Field, C. R. 1970. A study of the feeding habits of the hippopotamus in the Queen Elizabeth National Park, Uganda, with some management implications. *Zool. Afr.* 5:71–86.

Fisher, R. A. 1958. *The genetical theory of natural selection.* 2nd ed. New York: Dover Press.

Fogg, G. E. 1966. The extracellular products of algae. *Ocean. Mar. Biol.* 4:195–212.

______. 1971. Extracellular products of algae in fresh water. *Ergeb. Limnol.* 5:1–25.

______. 1975. *Algal cultures and phytoplankton ecology.* 2nd. ed. Madison: Univ. of Wisconsin Press.

Fox, L. R. 1975. Cannibalism in natural populations. *Ann. Rev. Ecol. Syst.* 6:87–107.

Fujita, H. and Utida, S. 1953. The effect of population density on the growth of an animal population. *Ecology* 34:488–98.

Gadgil, M. 1975. Evolution of social behavior through interpopulation selection. *Proc. Nat. Acad. Sci.* 72:1199–1201.

Gauthreaux, S. A. Jr. 1978. The ecological significance of behavioral dominance. In *Perspectives in ethology,* Vol. 3, eds. P. Klopfer and G. Bateson, pp. 17–54. New York: Plenum Press.

Ghiselin, M. T. 1974. *The economy of nature and the evolution of sex.* Berkeley: Univ. Cal. Press.

Gibson, T. E. 1953. The effect of repeated anthelmintic treatment with phenothiazine on the faecal egg counts of housed horses, with some observations on the life cycle of *Trichonema* spp. in the horse. *J. Helminthol* 26:29–40.

Gill, D. E. 1974. Intrinsic rate of increase, saturation density and competitive ability II. The evolution of competitive ability. *Am Nat.* 108:103–16.

Gilpin, M. E. 1975. The theory of group selection in predator-prey communities. Princeton: Princeton Univ. Press.

Glynn, P. W. 1976. Some physical and biological determinants of coral community structure in the eastern Pacific. *Ecol. Mon.* 46:431–56.

Goldsmith, W. M. 1916. Field notes on the distribution and life habits of the tiger beetles (Cicindelidae) of Indiana. *Indiana Acad. Sci. Prog.* 26:447–55.

Golley, F. B.; Ryszkowski, L.; and Sokur, J. T. 1975. The role of small mammals in temperate forests, grasslands and cultivated fields. In *Small mammals, their productivity and population dynamics*, IBP Handbook No. 5, eds. F. B. Golley, K. Petrusewics, and L. Ryszkowski, pp. 223–241. Cambridge: Cambridge Univ. Press.

Goodman, D. 1975. The theory of diversity-stability relationships in ecology. *Quart. Rev. Biol.* 50:237–66.

Graham, K. 1967. Fungal-insect mutualism in trees and timber. *Ann. Rev. Ent.* 12:105–26.

Grygierek, E. 1962. The effect of increasing the carp fry density on crustacean plankton. *Rocz. Nauk. Roln. Ser. B* 81(2):189–210.

Grygierek, E.; Hillbricht-Ilkowska, A.; and Spodniewska, I. 1966. The effect of fish on plankton communities in ponds. *Verh. Int. Ver. Limnol.* 16:1359–66.

Gunner, H. B. and Zuckerman, B. M. 1968. Degradation of Diazinon by synergistic microbial action. *Nature* 217:1183–84.

Haldane, J. B. S. 1932. *The causes of evolution.* London: Longmans, Green. (Reprinted as paperback, Cornell Univ. Ithaca, 1966.)

Hamilton, W.D. 1963. The evolution of altruistic behavior. *Am. Nat.* 97:354–56.

———. 1964. The genetical evolution of social behavior, I and II. *J. Theor. Biol.* 7:1–52.

———. 1970. Selfish and spiteful behavior in an evolutionary model. *Nature* 228(5277):1218–20.

———. 1971. Selection of selfish and altruistic behavior in some extreme models. In *Man and beast: comparative social behavior*, eds. J. F. Eisenberg and W. S. Dillon, pp. 57–91. Washington, D.C.: Smithsonian Institution Press.

———. 1975. Innate social aptitudes in man, an approach from evolutionary genetics. In *Biosocial anthropology*, ed. R. Fox, pp. 133–55. London: Malaby Press.

Hargrave, B. T. 1970. The effect of a deposit-feeding amphipod on the metabolism of benthic microflora. *Limn. and Ocean.* 15:21–30.

Hassel, M. P. 1971. Mutual interference between searching insect parasites. *J. Anim. Ecol.* 40:473–86.

Hendler, G. and Franz, D. R. 1971. Population dynamics and life history of *Crepidula convexa* Say (Gastropoda:Prosobranchia) in Delaware Bay. *Biol. Bull.* 141:514–26.

Hughes, R. D. ; Woolcock L. T.; and Ferrar, P. 1974. The selection of natural enemies for the biological control of the Australian bushfly. *J. Appl. Ecol.* 11:483–88.

Istock, C. A. 1967. The evolution of complex life cycle phenomena; an ecological perspective. *Evolution* 21:592–605.

Istock, C. A. and Weisburg, W. (m.s.) Genetic variation and population structure in *Wyeomyia* mosquitos.

Johannes, R. E. et al. 1972. The metabolism of some coral reef communities—a team study of nutrient and energy flux at Eniwetok. *BioScience* 22:541–43.

Johnson, C. G. 1969. *Migration and dispersal of insects by flight.* London: Methuen.

Johnston, R. 1964. Sea water, the natural medium of phytoplankton II. Trace metals and chelation, and general discussion. *J. Mar. Biol. Ass. U.K.* 44:87–109.

Joshi, M. M. and Hollis, J. P. 1977. Interaction of *Beggiatoa* and rice plant: detoxification of hydrogen sulfide in the rice rhizosphere. *Science* 195:179–80.

Katznelson, H. and Cole, S. E. 1965. Production of gibberillin-like substances by bacteria and actinomycetes. *Can. J. Microbiol.* 11:733.

Kennedy, C. R. 1974. Population biology of parasites with special reference to the effect of ecosystem changes due to human activity. *Proc. First. Int. Cong. Ecol.* 316–20.

King, C. E. and Dawson, P. S. 1972. Population biology and the Tribolium model. *Evol. Biol.* 5:133–227.

______. 1973. Habitat selection by flour beetles in complex environments. *Phys. Zool.* 46:297–311.

King, J. A. 1973. The ecology of aggressive behavior. *Ann. Rev. Ecol. Syst.* 4:117–38.

King, T. L. 1977. The plant ecology of ant hills in calcareous grasslands. I. Patterns of species in relation to ant hills in southern England. II. Succession on the mounds. III. Factors affecting the population sizes of selected species. *J. Ecol.* 65:235–56, 257–78, 279–315.

Kinn, D. N. 1971. The life cycle and behavior of *Cerocoleipus coelonotus* (Acarina:Mesostigmata), including a survey of phoretic mite associates of California Scolytidae. *U. Cal. Pub. Ent.* 65:1–62.

Kinn, D. N. and Witcosky, J. J. 1977. The life cycle and behavior of *Macrocheles bondreauxi* Krantz. *Z. Ang. Ent.* 84:136–144.

Krantz, G. W and Mellott, J. L. 1972. Studies on phoretic specificity in *Macrocheles mycotrupetes* and *M. peltotrupetes*—associates of geotrupine Scarabaeidae. *Acarologia* 14:317–44.

Krebs, C. J.; Wingate, I.; Ledus, J.; Redfield, J. A.; Taitt, M.; and Hilborn, R. 1976. *Microtus* population biology: dispersal in fluctuating populations of *M. townsendii*. *Can. J. Zool.* 54:79–95.

Lack, D. 1966. *Population studies of birds.* Oxford: Oxford Univ. Press.

Lamarra, V. 1975. Digestive activities of carp as a major contributor to the nutrient loading of lakes. *Verh. Int. Ver. Linmol.* 19:2961–68.

Larochelle, A. 1974. The food of Cicindelidae of the world. *Cicindela* 6:21–43.

Leston, D. 1970. Entomology of the cocoa farm. *Ann. Rev. Ent.* 15:273–94.

______. 1973. The ant mosaic-tropical tree crops and the limiting of pests and diseases. *PANS (Pest Artic. News Summ.)* 19:311–41.

Levene, H. 1953. Genetic equilibrium when more than one ecological niche is available. *Am. Nat.* 87:331–33.

Levin, B. R. and Kilmer, W. C. 1974. Interdemic selection and the evolution of altruism: a computer simulation study. *Evolution* 28:527–45.

Levin, S. A. 1976. Population dynamic models in heterogeneous environments. *Ann. Rev. Ecol. Syst.* 7:287–310.

Levin, S. A. and Udovic, J. D. 1977. A mathematical model of coevolving populations. *Am. Nat.* 111:657–75.

Levins, R. 1970. Extinction. In *Some mathematical questions in biology, lectures on mathematics in the life sciences,* ed. M. Gerstenhaber, pp. 77–107. Providence: Am. Math Soc.

______. 1974. Qualitative analysis of partially specified systems. *Ann. N.Y. Acad. Sci.* 231:123–38.

______. 1975. Evolution in communities near equilibrium. In *Ecology and evolution of communities,* eds. M. J. Cody Press. pp. 16–50. and J. M. Diamond, pp. 16–50. Cambridge: Harvard Univ. Press.

Lewis, J.R. and Bowman, R. S. 1975. Local habitat induced variation in the population dynamics of *Pattella vulgata* L. *J. Exp. Mar. Biol. Ecol.* 17:165–203.

Lin, N. and Michener, C. D. 1972. Evolution of sociality in insects. *Quart. Rev. Biol.* 47:131–59.

Lindquist, E. E. 1969a. Mites and the regulation of bark beetle populations. *Proc. 2nd. Int. Cong. Acar.* (Sutton, Bonington, England, 1967; Akad. Kiado, Budapest) 389–99.

______. 1969b. Review of Holarctic tarsonemid mites (Acarina: Prostigmata) parasitizing eggs of ipine bark beetles. *Ent. Soc. Can. Mem.* 60:111.

______. 1975. Associations between mites and other arthropods in forest floor habitats. *Can. Ent.* 107:425–37.

Lloyd, M. 1967. Mean crowding. *J. Anim. Ecol.* 36:1–30.

Lockwood, J. L. 1977. Fungistasis in soil. *Biol. Rev.* 52:1–45.

Losey, G. S. Jr. 1974. Cleaning symbiosis in Puerto Rico with comparison to the tropical Pacific. *Copeia* 1974:960–70.

Lucas, C. E. 1961. Interrelationships between aquatic organisms mediated by external metabolites. In *Oceanography,* ed. M. Sears, pp. 499–517. Washington: A.A.A.S.

Lugo, A. E., Franworth, E. G.; Pool, D.; Jerez, P.; and Kaufman, G. 1973. The impact of the leaf cutter ant *Atta colombica* on the energy flow of a tropical rain forest. *Ecology* 54:1292–1301.

MacNulty, B. 1971. An introduction to the study of Acari-Insecta associations. *Proc. Trans. Ent. Nat. Hist. Soc.* 4:46–70.

Margalef, R. 1968. *Perspectives in Ecological Theory.* Chicago: Univ. of Chicago Press.

Matessi, C. and Jayakar, S. D. 1973. A model for the evolution of altruistic behavior. *Genetics* 74:S174.

______. 1976. Conditions for the evolution of altruism under Darwinian selection. *Theor. Popl. Biol.* 9:360–87.

Matthewman, W. G. and Pielou, D. P. 1971. Arthropods inhabiting the sporophores of *Fomes fomentarius* (Polyporaceae) in Gatineau Park, Quebec. *Can. Ent.* 103:775–847.

Mattson, W. J. and Addy, N. D. 1975. Phytophagous insects as regulators of forest primary production. *Science* 190:515.

May, R. M. 1975a. *Stability and complexity in model ecosystems.* 2nd ed. Princeton: Princeton Univ. Press.

May, R. M., ed. 1976. *Theroretical ecology.* Oxford: Blackwell.

Maynard Smith, J. 1974. The theory of games and the evolution of animal conflicts. *J. Theor. Bio.* 47:209–21.

______. 1976. Group selection. *Quart. Rev. Biol.* 51:277–283.

Maynard Smith, J. and Parker, G. A. 1976. The logic of asymmetric contests. *Anim. Beh.* 24:159–75.

Maynard Smith, J. and Price, G. R. 1973. The logic of animal conflict. *Nature* 246:15–18.

Mayr, E. 1963. *Animal species and evolution.* Cambridge: Belknap.

McAllister, C. D. 1969. Aspects of estimating zooplankton production, and the adaptive value of vertical migration. *J. Fish. Res. Bd. Canada* 20:685–727.

McCracken, G. F. and Bradbury, J. W. 1977. Paternity and genetic heterogeneity in the polygynous bat. *Phyllostomus hastatus. Science* 198:303–6.

McLachlan, A. J. 1975. The role of aquatic macrophytes in the recovery of the benthic fauna of a tropical lake after a dry phase. *Limn. and Ocean.* 20:54–64.

McNaughton, S. J. 1975. Serengeti migratory wildebeest: facilitation of energy flow by grazing. *Science* 191:92–93.

______. 1977. Diversity and stability of ecological communities: a comment on the role of empiricism in ecology. *Am. Nat.* 111:515–25.

______. 1978. Serengeti Ungulates: feeding selectivity influences the effectiveness of plant defense guilds. *Science* 199:806–7.

Mertz, D. B. and Wade, M. J. 1976. The prudent prey and the prudent predator. *Am. Nat.* 489–96.

Meyer, J. S.; Tsuchiya, H. M.; and Fredrickson, A. G. 1975. Dynamics of mixed populations having complementary metabolism. *Biotechnol. Bioeng.* 17:1065–81.

Michel, J. F. 1963. The phenomena of host resistance and the course of infection of *Ostertagia osteragi* in calves. *Parasitology* 53:63–84.

Michener, C. D. and Brothers, D. J. 1974. Were workers of eusocial Hymenoptera initially altruistic or oppressed? *PNAS USA* 71(3):671–74.

Miller, C. B. 1970. Some environmental consequences of vertical migration in marine zooplankton. *Limn. and Oceanogr.* 15:727–42.

Milne, L. J. and Milne, M. 1976. The social behavior of burying beetles. *Sci. Amer. V.* 235:84–90.

Moser, J. C. 1975. Mite predators of the southern pine beetle. *Ann. Ent. Soc. Am.* 68:1113–16.

______. 1976. Surveying mites (Acarina) phoretic on the southern pine beetle (Coleoptera:Scolytidae) with sticky traps. *Can. Ent.* 108:809–13.

Moser, J. C. and Roton, L. M. 1971. Mites associated with southern pine bark beetles in Allen Parish, Louisiana. *Can. Ent.* 103:1175–98.

Moser, J. C.; Cross, E. A.; and Roton, L. M. 1971. Biology of *Pyemotes parviscolyti* (Acarina:Pyemotidae). *Entomophaga* 16:367–79.

Moser, J. C.; Thatcher, R. C.; and Pickard, L. S. 1971. Relative abundance of southern pine beetle associates in east Texas. *Ann. Ent. Soc. Am.* 64:72–77.

Muscatine, L. and Porter, J. W. 1977. Reef corals: mutualistic symbioses adapted to nutrient poor environments. *BioScience* 27:454–60.

Naumov, N. P. 1975. The role of rodents in ecosystems of the northern deserts of Eurasia. In *Small mammals, their productivity and population dynamics*, IBP Handbook No. 5, eds. F. B. Golley, K. Petrusewica, and L. Ryszkowski, pp. 299–309. Cambridge: Cambridge Univ. Press.

Neill, W. E. 1975. Experimental studies of microcrustacean competition, community composition, and efficiencies of resource utilization. *Ecology* 56:809–26.

Neuweiler, G. 1969. Verhaltensbeobachtungen an einer indischen Flughundkolonie (*Pteropus g. giganteus* Brunn). *Zeit. fur Tierpsychol.* 26:166–99.

Odum, E. P. 1969. The strategy of ecosystem development. *Science* 164: 262–70.

______. 1971. *Fundamentals of ecology.* 3rd ed. Philadelphia: W. B. Saunders Co.

Olsen, O. W. 1974. *Animal parasites: their life cycles and ecology.* Baltimore: University Park Press.

Pahlsson, L. 1974. Influence of vegetation of microclimate and soil moisture on a Scanian hill. *Oikos* 25(2):176–86.

Park, T.; Mertz, D. B.; Mertz, B.; Grodzinski, W.; and Prus, T. 1965. Cannibalistic predation in populations of flour beetles. *Physiol. Zool.* 38:259–321.

Parker, G. 1974. Assessment strategy and the evolution of animal conflicts. *Jour. Theor. Biol.* 47:223–43.

Pearson, L. C. 1965. Primary production in grazed and ungrazed desert communities of eastern Idaho. *Ecology* 46:278–85.

Pianka, E. 1974. *Evolutionary Ecology.* New York: Harper & Row.

Pielou, D. P. and Verma, A. N. 1968. The arthropod fauna associated with the birch bracket fungus *Polyporus betulinus* in eastern Canada. *Can. Ent.* 100:1179–99.

Pimentel, D. 1961. Animal population regulation by the genetic feedback mechanism. *Am. Nat.* 95:65–79.

Price, P. W. 1970. Characteristics permitting coexistence among parasitoids of a sawfly in Quebec. *Ecol.* 51:445–54.

Rafes, P. M. 1971. Pests and the damage which they cause to forests. In *Productivity of forest ecosystems, Proc. Brussels Symp. 1967*, pp. 357–367. Paris: UNESCO, Ecology and Conservation, 4.

Rapp, W. F. Jr. and Emil, C. 1965. Mosquito production in a eutrophic sewage stabilization lagoon. *J. Water Poll. Contr. Fed.* 37:867–70.

Redfield, J. A. 1975. Comparative demography of increasing and stable populations of blue grouse (*Dandrogapus obscurus*). *Can. J. Zool.* 53:1–11.

Regenfuss, H. 1972. Uber die Einnischung synhospitaler Parasitenarten auf dem Wirtskorper. Untersuchungen an ektoparasitischen Milben (Podapolipidae) auf laufkafern (Carabidae). *Z. Zool. Systematik. u. Evolutionsforschung* 10:44–65.

Rhoads, D. C. and Young, D. K. 1970. The influence of deposit-feeding benthos on bottom stability and community trophic structure. *J. Mar. Res.* 28:150–78.

______. 1971. Animal-sediment relations in Cape Cod Bay, Mass. II. Reworking by *Molpadia volitica* (Holothuroidea). *Mar. Bol.* 11:225–61.

Ricklefs, R. E. 1972. *Ecology*. Portland: Chiron Press.

Rohwer, S. 1975. The social significance of avian winter plumage variability. *Evolution* 29:593–610.

______. 1977. Status signaling in Harris sparrows, some experiments in deception. *Behavior* 61:107–29.

Rohwer, S.; Fretwell, S. D.; and Tuckfield, R. C., 1976. Distress screams as a measure of kinship in birds. *Am. Midl. Nat.* 96:418–30.

Rohwer, S. and Rohwer, F. C. 1978. Status signaling in Harris sparrows, experimental deceptions achieved. *Animal Behavior* 26:1012–1022.

Rohwer, S. and Winfield, J. C. (m.s.). Social dominance, plasma LH and steroids in a winter population of sparrows: a field study.

Room, P. M. 1971. The relative distributions of ant species in Ghana's cocoa farms. *J. Anim. Ecol.* 40:735–51.

______. 1973. Control by ants of pest situations in tropical tree crops; a strategy for research and development. *Papua New Guinea Agricultural Journal* 24:98–103.

______. 1975. Relative distributions of ant species in cocoa plantations in Papua New Guinea. *J. Appl. Ecol.* 12:47–61.

Root, R. B. 1975. Some consequences of ecosystem texture. In *Ecosystem analysis and prediction*, ed. S. A. Levin, pp. 83–97. Philadelphia: Soc. Ind. App. Math.

Roper, M. M. and Marshall, K. C. 1974. Modification of the interaction between *Escherichia coli* and bacteriophage in saline sediment. *Microb. Ecol.* 1:1–13.

Rosenzwieg, M. L. 1973. Evolution of the predator isocline. *Evolution* 27:84–94.

Roughgarden, J. 1975. Evolution of marine symbiosis—a simple cost benefit model. *Ecology* 56:1201–08.

______. 1976. Resource partitioning among competing species—a coevolutionary approach. *Theor. Pop. Biol.* 9:388–424.

______. 1977. Coevolution in ecological systems: results from "loop analysis" for purely density-dependent coevolution. In *Measuring selection in natural populations,* eds. F. B. Chistiensen and P. M. Fenchel, pp. 501–18. Berlin: Springer Verlag.

______. 1978. *Theory of population genetics and evolutionary ecology, an introduction.* New York: Macmillan.

Rowell, T. E. 1974. The concept of social dominance. *Beh. Biol.* 11:131–54.

Russo, G. 1926. Contributo alla conescenza degli Scolytidi. Studio morfobiologico del *Chaetoptelius vestitus* (Muls. e. Rey) Fuchs e dei suoi simbionti. *Boll. Lab. Zool. Gen. Agr. R. Scu. Sup. Agric. Portici.* 19:101–260.

______. 1938. VI. Contributo alla conoscenze dei Coleotteri Scolitide. Fleotribo: *Phloeotribus* scarabaeoides (Bern.) Fauv. Parte Seconda. Biografia, simbionti, danni e lotta. *Boll. R. Lab. Ent. Agr. Portici.* 2:3–420.

Schad, G. A. 1977. The role of arrested development in the regulation of nematode populations. In *Regulation of parasite populations,* ed. G. W. Esch, pp. 11–168. New York: Academic Press.

Schoener, T. W. 1976. Alternatives to Lotka-Volterra competitions: models of intermediate complexity. *Theor. Pop. Biol.* 10(3):309–34.

Shubik, M. 1975. *The uses and methods of gaming.* New York: Elsevier.

Skaife, S. H. 1952. The yellow-banded carpenter bee, *Mesotrichia caffra* Linn. and its symbiotic mite *Dinogamasus braunsi* Vitzthum. *J. Ent. Soc. South. Afr.* 15:63–76.

Slatkin, M. and Wade, M. J. 1978. Group selection on a quantitative character. *Proc. Nat. Acad. Sci.* 75:3531–34.

Slobodkin, L. B. 1961. Growth and regulation of animal populations. New York: Holt, Rinehart and Winston.

______. 1968. How to be a predator. *Am. Zool.* 8:43–51.

______. 1974. Prudent predation does not require group selection. *Am. Nat.* 108:665–78.

Slobodkin, L. B. and Richman, S. 1956. The effect of removal of fixed percentages of the newborn on size and variability in populations of *Daphnia pulicaria. Limn. and Oceanogr.* 2:209–37.

Smith, Adam. 1776. The wealth of nations. Reprint. New York: Random House, 1937. (Random House edition 1937).

Sokoloff, A. 1955. Competition between sibling species of the *Pseudoobscura* subgroup of *Drosophila. Ecol. Monog.* 25:387–409.

Spedding, C. R. W. 1971. Grassland ecology. Oxford: Oxford Univ. Press.

Springett, B. P. 1968. Aspects of the relationship between burying beetles *Necrophorus* spp. and the mite *Poecilochirus necrophori. J. Anim. Ecol.* 37:417–24.

Tahvanainen, J. O. and Root, R. B. 1972. The influence of vegetational diversity on the population ecology of a specialized herbivore *Phylotreta cruciferae. Oecologia* 10:321–46.

Tanner, J. T. 1966. Effects of population density on growth rates of animal populations. *Ecology* 47:733–45.

Taylor, W. E. and Bardner, R. 1968. Effects of feeding by larvae of *Phaedon cochlearae* (F.) and *Plutella maculipennis* (Curt.) on the yield of radish and turnip plants. *Ann. Appl. Biol.* 62:249–54.

Thalenhorst, W. 1958. Grundzüge der Populationsdynamik des grossen Fichtenborkenkäfers *Ips typographus* L. SchrReihe Forstl. Fak. Univ. Göttingen 21.

Thomas, W. A. and Grigal, D. F. 1976. Phosphorus conservation by evergreenness of mountain laurel. *Oikos* 27:19–26.

Trivers, R. 1971. The evolution of reciprocal altruism. *Quart. Rev. Biol.* 46:35–57.

Valiela, I. 1969. An experimental study of the mortality factors of larval *Musca autumnalis* De Geer. *Ecol. Monog.* 39:199–225.

______. 1974. Composition, food webs and population limitation in dung arthropod communities during invasion and succession. *Am. Mid. Nat.* 92:370–85.

Verner, J. 1977. On the adaptive significance of territoriality. *Am. Nat.* 111:769–75.

Vickery P. J. 1972. Grazing and net primary production of a temperate grassland. *J. Appl. Ecol.* 9:307.

Wade, C. F. and Rodriguez, T. G. 1961. Life history of *Macrocheles muscaedomesticae,* a predator of the housefly. *Ann. Ent. Soc. Am.* 54:776–81.

Wade, M. J. 1976. Group selection among laboratory populations of *Tribolium. Proc. Nat. Acad. Sci.* 73:4604–7.

______. 1977. Experimental study of group selection. *Evolution* 31:134–53.

______. 1978. A critical review of the models of group selection. *Quart. Rev. Biol.* 53:101–14.

Wade, M. T. 1978a. The selfish gene (book review). *Evolution* 32:220–21.

Wahlund, S. 1928. Zuzammensetzung von populationen und korrelationserscheinungen von standpunkt der vererbungslehre aus betrachtet. *Hereditas* 11:65–106.

Ward, P. 1965. Feeding ecology of the black-faced Dioch *Quelea quelea* in Nigeria. *Ibis.* 107:173–214.

Waterhouse, D. F. 1974. The biological control of dung. *Scientific American* 230:101–109.

Welsh, B. L. 1975. The role of grass shrimp, *Palaemonetes pugio,* in a tidal marsh ecosystem. *Ecol.* 56:513–30.

West Eberhard, M. J. 1975. The evolution of social behavior by kin selection. *Quart. Rev. Biol.* 50:1–33.

Whittaker, R. H. and Feeny, P. P. 1971. Allelochemics: chemical interactions between species. *Science* 171:757–70.

Whittaker, R. H. and Levin, S. 1977. The role of mosaic phenomena in natural communities. *Theor. Pop. Biol.* 12:117–39.

Wiens, J. A. 1976. Population responses to patchy environments. *Ann. Rev. Ecol. Syst.* 7:81–120.

Willams, G. C. 1966. *Adaptation and natural selection: a critique of some current evolutionary thought.* Princeton: Princeton Univ. Press.

———. 1975. *Sex and evolution.* Princeton: Princeton Univ. Press.

Wilson, D. S. 1974. Prey capture and competition in the ant lion. *Biotropica* 6:187–93.

———. 1975a. A theory of group selection. *Proc. Nat. Acad. Sci.* 72:143–46.

———. 1975b. The adequacy of body size as a niche difference. *Am. Nat.* 109:769–84.

———. 1976. Evolution on the level of communities. *Science* 192:1358–60.

———. 1977a. Structured demes and the evolution of group advantageous traits. *Am. Nat.* 111:157–185.

———. 1977b. How nepotistic is the brain worm? *Behav. Ecol. Sociobiol.* 2:421–425.

———. 1978. Prudent predation—a field test involving three species of tiger beetles. *Oikos* 31:128–36.

———. 1979. Structured demes and trait-group variation. *Am. Nat.* 113:606–10.

Wilson, D. S.; Leighton, M.; and Leighton, D. 1979. Interference competition in a tropical ripple bug. *Biotropica* 10:302–06.

Wilson, E. O. 1971. *The insect societies.* Cambridge: Belknap Press.

———. 1973. Group selection and its significance for ecology. *BioScience* 23(11):631–38.

———. 1975. *Sociobiology: the new synthesis.* Cambridge: Harvard Univ. Press.

Witkamp, M. 1971. Soils as components of ecosystems. *Ann. Rev. Ecol. Syst.* 2:85–110.

Wright, S. 1938. Size of population and breeding structure in relation to evolution. *Science* 87:430–31.

———. 1945. Tempo and mode in evolution: a critical review. *Ecology* 26(4):415–19.

Wynne-Edwards, V. C. 1962. *Animal dispersion in relation to social behavior.* Edinburgh: Oliver and Boyd.

Yeoh, H. T.; Bungay, H. R.; and Krieg, N. R. 1968. A microbial interaction involving combined mutualism and inhibition. *Can. J. Microbiol.* 14:491–92.

Zinke, P. J. 1962. The pattern of individual forest trees on soil properties. *Ecology* 43:130–33.

Index